国家级实验教学示范中心联席会
计算机学科组规划教材

WinForm程序设计与实践

第2版

廉龙颖 主 编
王希斌 游海晖 赵艳芹 副主编

清华大学出版社
北京

内 容 简 介

本书分为3部分，共9章。第1~3章为基础知识部分，主要介绍C#的基本语法和面向对象程序设计方法；第4~6章为WinForm程序设计部分，主要介绍基于.NET平台的Windows程序开发，包括WinForm基础、输入与输出、数据访问技术等；第7~9章为实践项目部分，主要以初级项目研究生科研管理系统、中级项目酒店管理系统和高级项目学生选课管理系统为案例，分别从易到难完整地介绍WinForm项目的设计过程。

本书是在作者多年讲授.NET程序设计课程的讲义基础上整理而成的，包含多年实际经验。本书力求内容组织合理，难易适当，叙述简洁流畅，语言通俗易懂，示例简短精练，能够使学生轻松、愉快地掌握WinForm程序设计方法和技巧。本书可作为全国高等学校计算机相关专业的教材，也可作为培训机构学生学习的参考用书。

版权所有，侵权必究。举报：010-62782989，beiqinquan@tup.tsinghua.edu.cn。

图书在版编目（CIP）数据

WinForm程序设计与实践 / 廉龙颖主编. -- 2版. -- 北京：清华大学出版社，2025.3.
（国家级实验教学示范中心联席会计算机学科组规划教材）. -- ISBN 978-7-302-68635-4

Ⅰ. TP316.7

中国国家版本馆CIP数据核字第2025D4V610号

责任编辑：陈景辉
封面设计：刘　键
责任校对：李建庄
责任印制：刘海龙

出版发行：清华大学出版社
　　　网　　址：https://www.tup.com.cn，https://www.wqxuetang.com
　　　地　　址：北京清华大学学研大厦A座　　邮　　编：100084
　　　社 总 机：010-83470000　　　　　　　　邮　　购：010-62786544
　　　投稿与读者服务：010-62776969，c-service@tup.tsinghua.edu.cn
　　　质 量 反 馈：010-62772015，zhiliang@tup.tsinghua.edu.cn
　　　课 件 下 载：https://www.tup.com.cn，010-83470236
印 装 者：北京瑞禾彩色印刷有限公司
经　　销：全国新华书店
开　　本：185mm×260mm　　印　张：19.75　　字　数：482千字
版　　次：2019年3月第1版　2025年5月第2版　印　次：2025年5月第1次印刷
印　　数：1~1500
定　　价：59.90元

产品编号：107238-01

前 言

.NET 框架是微软公司在 2000 年专业开发者会议上提出的发展中的开发平台，这是一个革命性的应用程序开发平台。在该平台中，C#作为微软公司面向对象的下一代应用平台的核心语言，能够让开发人员在.NET 平台上快速开发应用程序。

无论高校还是 IT 培训学校，目前都将 .NET 作为教学内容之一，这对于培养学生的计算机程序设计能力具有非常重要的意义。在已开设的 .NET 相关课程中，主要分为 WinForm 程序设计和 Web 程序设计，虽然.NET 教材较多，但大部分都是以 Web 程序设计为主，专门针对基于.NET 平台的 WinForm 程序设计的教材非常少。因此，我们编写了本书。

本书特色

（1）知识结构完整。根据循序渐进的认知规律设计编写内容和顺序。

（2）示例简短精练。所有知识点都设计了一个针对性强的示例，所有示例都通过 Visual Studio .NET 2022 调试，并给出了运行结果，其中部分复杂的实例还有详细的分析，以帮助读者理解。

（3）习题丰富多样。全书各章配备了丰富的标准化习题，便于教师教学和考试。

配套资源

为便于教与学，本书配有源代码、教学课件、教学大纲、教案、教学进度表、习题题库、期末试卷及答案。

（1）获取源代码、全书网址方式：先刮开并用手机版微信App扫描本书封底的文泉云盘防盗码，授权后再扫描下方二维码，即可获取。

源代码　　全书网址

（2）其他配套资源可以扫描本书封底的"书圈"二维码，关注后回复本书书号，即可下载。

读者对象

本书可作为全国高等学校计算机相关专业的教材，也可作为培训机构学生学习的参考用书。

通过本书的学习，可以使读者掌握 C#语言基础、ADO.NET 数据库访问技术及开发 .NET 程序的基础知识和基本方法，对 WinForm 程序设计有一个全面的认识，能够独立开发各类 WinForm 应用程序，并为后期学习基于.NET 的 Web 程序设计奠定基础。

本书作为教材使用时，建议在实验室授课，采用课堂教学与实验教学相结合的方式进行，建议授课32学时，课程设计2周。各章学时建议分配如下，教师可以根据实际教学情况进行调整。

章	内　　容	学时
第1章	.NET简介及其开发环境	2
第2章	C#语言基础	6
第3章	面向对象程序设计	6
第4章	WinForm基础	6
第5章	输入与输出	4
第6章	数据访问技术	8
第7~9章	实践项目——研究生科研管理系统/酒店管理系统/学生选课管理系统	2周

感谢为本书提出建议的所有老师和学生。本书第1~4章由廉龙颖编写，第5~6章由王希斌编写，第7、9章由游海晖编写，第8章由赵艳芹编写。

由于编者水平有限，编写时间仓促，书中难免存在不足之处，希望读者批评指正。

编　者

2025年1月

目 录

第 1 章 .NET 简介及其开发环境 1

   1.1 .NET 简介 1
      1.1.1 .NET Framework 2
      1.1.2 C# 2
      1.1.3 Visual Studio 3
   1.2 搭建开发环境 4
   1.3 编程初试 5
      1.3.1 编写 HelloWorld 程序 5
      1.3.2 使用 Visual Studio 的技巧 10
   1.4 .NET 基本概述 12
      1.4.1 解决方案与项目 12
      1.4.2 命名空间 12
      1.4.3 Main()方法 13
      1.4.4 代码注释 14
      1.4.5 程序调试 14

第 2 章 C#语言基础 18

   2.1 C#语法元素 18
   2.2 关键字与标识符 18
      2.2.1 关键字 18
      2.2.2 标识符 19
   2.3 数据类型 20
      2.3.1 简单值类型 21
      2.3.2 结构 23
      2.3.3 枚举 24
      2.3.4 Object 26
      2.3.5 类 27
      2.3.6 接口 28
      2.3.7 字符串 30
      2.3.8 数组 31
   2.4 常量与变量 35

2.4.1 常量 35
　　2.4.2 变量 35
　　2.4.3 变量的作用域 35
2.5 运算符 36
　　2.5.1 算术运算符 36
　　2.5.2 赋值运算符 37
　　2.5.3 比较运算符 39
　　2.5.4 逻辑运算符 40
　　2.5.5 运算符优先级 42
2.6 流程控制语句 42
　　2.6.1 选择结构语句 42
　　2.6.2 循环结构语句 49
　　2.6.3 跳转语句 55
2.7 数据类型转换 59
2.8 异常处理 62

第 3 章　面向对象程序设计 67

3.1 面向对象简介 67
3.2 类与对象 67
　　3.2.1 类的声明 68
　　3.2.2 对象的创建与使用 69
3.3 类的数据成员 69
3.4 方法 71
　　3.4.1 方法的定义与调用 71
　　3.4.2 方法的重载 73
　　3.4.3 方法的高级参数 74
3.5 构造方法 77
3.6 访问修饰符与 static 关键字 79
　　3.6.1 访问修饰符 79
　　3.6.2 static 关键字 79
3.7 面向对象的基本特征 83
　　3.7.1 封装 84
　　3.7.2 继承 85
　　3.7.3 多态 86

第 4 章　WinForm 基础 88

4.1 WinForm 简介 88
　　4.1.1 WinForm 程序的新建 88
　　4.1.2 WinForm 程序的文件结构 89

	4.1.3	窗体与控件	91
	4.1.4	属性与事件	93
4.2	WinForm 常用控件		95
	4.2.1	文本类控件	96
	4.2.2	选择类控件	101
	4.2.3	分组类控件	105
	4.2.4	其他控件	109
4.3	Windows 通用对话框		116
	4.3.1	消息对话框	116
	4.3.2	文件对话框	117
	4.3.3	普通对话框	118
4.4	Windows 窗体设计		121
	4.4.1	基于单文档的窗体设计	121
	4.4.2	基于多文档的窗体设计	121

第 5 章 输入与输出 125

5.1	概述		125
	5.1.1	文件与流	125
	5.1.2	System.IO 命名空间	125
5.2	目录操作		126
	5.2.1	Directory 类	126
	5.2.2	DirectoryInfo 类	127
5.3	文件操作		130
	5.3.1	File 类	130
	5.3.2	FileInfo 类	131
5.4	文件读写		133
	5.4.1	读写文本文件	133
	5.4.2	读写二进制文件	135

第 6 章 数据访问技术 139

6.1	数据库基础		139
	6.1.1	数据库的基本概念	139
	6.1.2	数据库访问过程	140
6.2	ADO.NET		141
	6.2.1	ADO.NET 概述	141
	6.2.2	ADO.NET 数据库访问步骤	142
6.3	ADO.NET 数据库访问操作		143
	6.3.1	使用 Connection 对象连接数据库	145
	6.3.2	使用 Command 对象执行数据库命令	146
	6.3.3	使用 DataAdapter 对象执行数据库命令	152

第 7 章 实践项目——研究生科研管理系统 ... 156

- 7.1 需求分析 ... 156
- 7.2 概要设计 ... 156
  - 7.2.1 架构设计 ... 156
  - 7.2.2 功能设计 ... 156
- 7.3 数据库设计 ... 157
- 7.4 实体模型设计 ... 158
- 7.5 数据访问层设计 ... 163
- 7.6 业务逻辑层设计 ... 165
- 7.7 表示层设计 ... 174
  - 7.7.1 系统主窗体设计 ... 175
  - 7.7.2 添加学生信息设计 ... 178
  - 7.7.3 维护学生信息设计 ... 180
  - 7.7.4 添加导师信息设计 ... 183
  - 7.7.5 维护导师信息设计 ... 185
  - 7.7.6 添加论文成果设计 ... 188
  - 7.7.7 维护论文成果设计 ... 190
  - 7.7.8 添加竞赛成果设计 ... 193
  - 7.7.9 维护竞赛成果设计 ... 196
  - 7.7.10 查询导师成果设计 ... 200
  - 7.7.11 查询学生成果设计 ... 201

第 8 章 实践项目——酒店管理系统 ... 203

- 8.1 需求分析 ... 203
- 8.2 概要设计 ... 203
  - 8.2.1 架构设计 ... 203
  - 8.2.2 功能设计 ... 204
- 8.3 数据库设计 ... 204
- 8.4 实体模型设计 ... 206
- 8.5 数据访问层设计 ... 212
- 8.6 业务逻辑层设计 ... 214
- 8.7 表示层设计 ... 222
  - 8.7.1 登录设计 ... 223
  - 8.7.2 系统主界面设计 ... 224
  - 8.7.3 添加新用户设计 ... 227
  - 8.7.4 修改/删除用户设计 ... 229
  - 8.7.5 入住登记设计 ... 231
  - 8.7.6 退房登记设计 ... 236

  8.7.7 住客信息查询设计 ········ 240
  8.7.8 客房信息查询设计 ········ 245
  8.7.9 帮助设计 ········ 247

第9章 实践项目——学生选课管理系统 ········ 249

9.1 需求分析 ········ 249
9.2 概要设计 ········ 250
  9.2.1 架构设计 ········ 250
  9.2.2 功能模块设计 ········ 251
9.3 数据库设计 ········ 251
9.4 实体模型设计 ········ 253
9.5 数据访问层设计 ········ 258
9.6 业务逻辑层设计 ········ 260
9.7 表示层设计 ········ 269
  9.7.1 登录设计 ········ 269
  9.7.2 系统前台主窗体设计 ········ 272
  9.7.3 维护个人密码设计 ········ 274
  9.7.4 选课设计 ········ 276
  9.7.5 查看课程表设计 ········ 279
  9.7.6 系统后台主窗体设计 ········ 281
  9.7.7 系统信息管理设计 ········ 283
  9.7.8 添加学生信息设计 ········ 284
  9.7.9 维护学生信息设计 ········ 288
  9.7.10 添加课程信息设计 ········ 292
  9.7.11 维护课程信息设计 ········ 296
  9.7.12 维护成绩信息设计 ········ 302

参考文献 ········ 306

第1章 .NET 简介及其开发环境

学习目标：
- 认识.NET 平台与 C#语言；
- 了解.NET、.NET Framework、Visual Studio 及 C#语言之间的关系；
- 掌握开发环境搭建过程；
- 掌握 HelloWorld 程序的编写；
- 理解和掌握.NET 程序运行与调试过程。

1.1 .NET 简介

.NET 平台是由微软公司推出的应用程序开发平台，用来构建和运行 Microsoft Windows 和 Web 应用程序。对.NET 可以从两个方面理解。首先，.NET 是一个开发平台。它对微软公司之前的主要开发平台进行了集成，提供了一套全新的 Windows 平台。例如，在.NET 平台下不仅可以进行 Visual Basic、C++程序的开发，还可以使用特别为.NET 平台开发的 C#语言进行编程。.NET 平台要做到的就是消除互连环境中不同软硬件以及服务的差异，使得不同设备和系统都可以相互通信，并使得不同的程序和服务之间都可以相互调用。其次，.NET 是一组规范。.NET 平台本身就基于一系列规范，其中有些规范是由微软公司以外的其他组织维护的。例如，定义 C#及 Visual Basic 语言的规范、定义数据交换格式的规范等。

.NET 平台的核心是.NET Framework，它为.NET 平台下应用程序的运行提供基本框架。.NET Framework 是微软公司推出的一套类库，被称为.NET 框架，此框架最大的优点是支持 C#语言。

Visual Studio 是目前最流行的.NET 应用程序集成开发环境。.NET 平台是建立在开放体系结构基础之上的，应用程序开发人员也可以使用其他开发工具。在.NET 架构中，.NET Framework、C#及 Visual Studio 三者之间关系如图 1-1 所示。

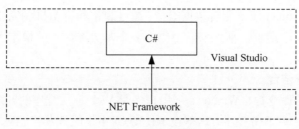

图 1-1 .NET 架构

.NET Framework 是运行.NET 应用程序的基础，而 Visual Studio 是开发.NET 应用程序的集成开发环境（Integrated Development Environment，IDE），Visual Studio 的运行要以.NET Framework 为基础。可以这样比喻三者的关系，.NET Framework 是.NET 程序运行的幕后操纵者；而 Visual Studio 是前台具体的操作者，如同.NET Framework 的外壳。

C#与.NET 的关系体现在两个方面：第一，C#的设计目的就是用来开发在.NET Framework 中运行的代码，因此，.NET Framework 是 C#程序的运行环境；第二，C#的编程库是.NET Framework 类库，即 C#的数据类型和操作类都来自于.NET 类库。

1.1.1　.NET Framework

.NET Framework 是微软公司为开发应用程序而创建的一个框架。使用.NET Framework 可以创建桌面应用程序、Web 应用程序、Web 服务及其他各种类型的应用程序。

.NET Framework 具有两个主要组件：公共语言运行库和.NET Framework 类库。

（1）公共语言运行库是.NET Framework 的基础。可以将运行库看作一个在执行时管理代码的代理，它提供内存管理、线程管理、远程处理等核心服务，并且还强制实施严格的类型安全以及可提高安全性和可靠性的其他形式的代码准确性。事实上，代码管理的概念是运行库的基本原则。以运行库为目标的代码称为托管代码，而不以运行库为目标的代码称为非托管代码。

（2）.NET Framework 类库是一个综合性的面向对象的可重用类型集合，可以使用它开发多种应用程序，这些应用程序包括传统的命令行或图形用户界面应用程序，也包括创新的应用程序，如 Web 窗体和 XML Web Services。

1.1.2　C#

C#是专门为.NET 设计的面向 Internet 和企业级应用的新一代编程语言，C#具有安全、稳定、简单、优雅等特点，是由 C 语言和 C++语言衍生的面向对象的编程语言，读作 C Sharp。C#语言的特点如下。

1．语法简洁

C#抛弃了 C 和 C++的指针，不允许代码直接操作内存。

C#自动计算数组或集合的长度，有效地避免了内存地址或数组下标越界的问题。

C#统一了对结构型、类及其成员的引用操作符，只有一个"."，使代码书写更简单。

C#没有全局方法，也没有全局变量，这使代码具有更好的可读性，也减少了因命名而造成的冲突。

2．完全面向对象设计

C#使用根类型 Object 统一所有数据类型，通过装箱和拆箱机制完成对象操作或数据类型转换；C#只允许单一继承，不允许一个类从多个基类派生，从根本上避免了类型定义的混乱问题。

3．与 Web 紧密结合

C#统一了传统的命令行、Windows 应用程序及 Web 应用程序的开发模式。同时，微软公司又推出了 WPF、WCF 等技术，这些技术使得 C#不仅能开发普通应用程序，还能在网络通信、动画制作、游戏开发、图像处理、多媒体应用、移动设备领域等发挥重要作用。

4. 完善的安全性与错误处理

在安全性方面，C#提供了完整的类型安全机制。例如，对象的成员变量由编译器负责初始化，而其他局部变量未经初始化则不允许使用，编译器也会进行自动检查并提示；CLR提供垃圾回收、类型安全检查、内部代码信任机制等，允许管理员或用户根据自己的 ID 配置安全等级，借助 CLR 这一特性，可以进一步确保 C#应用程序的安全性。

在错误处理上，C#借助 Visual Studio 的智能感知技术，可以消除在程序编写过程中的许多常见错误；C#还提供统一的异常类 Exception 管理程序在运行过程中产生的错误。

5. 良好的可扩展性

C#应用程序能跨语言、跨平台、跨互联网互相调用；C#语言允许自定义数据类型，以扩展元数据。

1.1.3 Visual Studio

Visual Studio 是目前最流行的 Windows 平台应用程序的集成开发环境，包括整个软件生命周期中所需要的大部分工具，简称 VS。

Visual Studio 的发展历史如表 1-1 所示。

表 1-1 Visual Studio 的发展历史

名 称	内部版本	发布日期	支持.NET Framework 版本
Visual Studio .NET 2002	7.0	2002-02-13	1.0
Visual Studio .NET 2003	7.1	2003-04-24	1.1
Visual Studio 2005	8.0	2005-11-07	2.0
Visual Studio 2008	9.0	2007-11-19	2.0、3.0、3.5
Visual Studio 2010	10.0	2010-04-12	2.0、3.0、3.5、4.0
Visual Studio 2012 RTM	11.0	2012-08-25	2.0、3.0、3.5、4.0、4.5、4.5.1、4.5.2、4.6、4.6.1、4.6.2
Visual Studio 2013	12.0	2013-10-17	2.0、3.0、3.5、4.0、4.5、4.5.1、4.5.2、4.6、4.6.1、4.6.2
Visual Studio 2015	14.0	2014-11-10	2.0、3.0、3.5、4.0、4.5、4.5.1、4.5.2、4.6、4.6.1、4.6.2
Visual Studio 2015 RTM	14.0	2015-07-21	2.0、3.0、3.5、4.0、4.5、4.5.1、4.5.2、4.6、4.6.1、4.6.2
Visual Studio 2017	15.0	2017-03-08	4.6、4.6.1、4.6.2、4.7、4.7.1、4.7.2、4.8
Visual Studio 2019	16.0	2019-04-02	4.6、4.6.1、4.6.2、4.7、4.7.1、4.7.2、4.8
Visual Studio 2022	17.1	2022-01-05	.NET Framework 全系列、.NET Core 全系列、.NET 5.0、.NET 6.0

Visual Studio 的特点如下所述。

（1）轻松创建简单、易用的应用程序，自定义窗口布局，为开发提供了一些便利。

（2）集成多种控件，这些控件涵盖了 Web 应用、数据库应用等领域，使开发工作更加简便、快速。

（3）高级的调试、配置、自动化和手工测试工具。

（4）代码编辑器支持代码彩色显示、智能感知、语法校对等功能。

（5）提供内置的可视化数据库工具，使开发数据库应用程序更加方便。

1.2　搭建开发环境

开发.NET 程序，首先要搭建开发环境，本书使用 Visual Studio 2022 作为开发工具。

.NET Framework 可在微软公司官网（网址详见前言二维码）下载 Community 2022 版本，下载后根据安装向导界面提示进行安装，安装后重新启动计算机。安装步骤如下所述。

（1）双击可执行文件 VisualstudioSetup.exe，启动如图 1-2 所示的 Visual Studio Installer 安装程序向导。

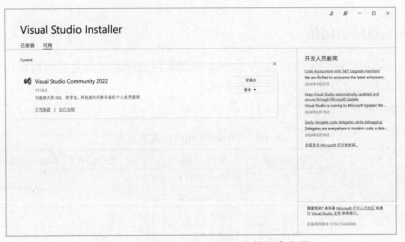

图 1-2　Visual Studio Installer 安装程序向导

（2）单击"安装"按钮，进入如图 1-3 所示的安装界面。选中".NET 桌面开发"等复选框，然后单击"安装"按钮开始安装。

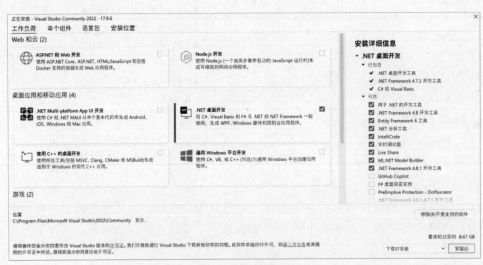

图 1-3　安装界面

（3）安装完成，如图1-4所示。

图1-4　安装完成

1.3　编程初试

1.3.1　编写HelloWorld程序

开发.NET项目的一般步骤是：创建解决方案，在解决方案中创建项目，然后在项目生成的代码页中编写代码后运行，查看运行结果。如果程序代码有错误，需要排错后再重新运行。下面以控制台项目HelloWorld为例介绍编程过程。

1. 新建解决方案

打开Visual Studio 2022，如图1-5所示，选择"创建新项目"选项，将打开"创建新项目"对话框，在搜索框中输入"解决方案"，选择"空白解决方案"选项，如图1-6所示。

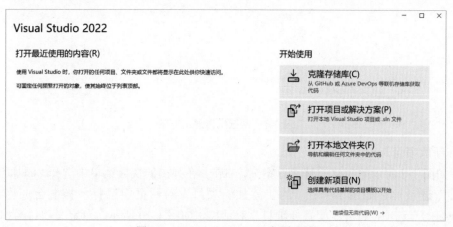

图1-5　Visual Studio 2022初始界面

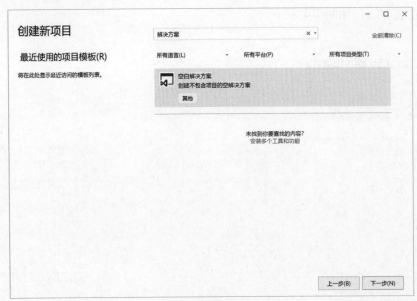

图 1-6　创建空白解决方案

在打开的"配置新项目"窗口中,自定义解决方案名称并选择保存位置(这里命名为 Solution1,位置为"C:\示例代码\chapter01\"),单击"创建"按钮,创建一个空白解决方案,如图 1-7 所示。

图 1-7　配置空白解决方案

2. 新建项目

在如图 1-8 所示的新建的解决方案上,右击,在弹出的快捷菜单中选择"添加"→"新建项目"选项。在如图 1-9 所示的"添加新项目"对话框中选择"控制台应用(.NET Framework)"选项,打开"配置新项目"对话框,如图 1-10 所示。自定义项目名称(这里命名为 HelloWorld),单击"创建"按钮,创建一个控制台项目。

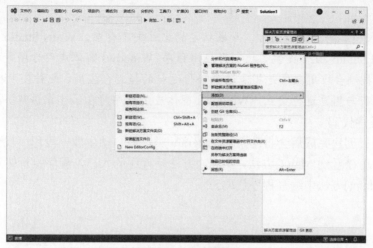

图 1-8 新建项目

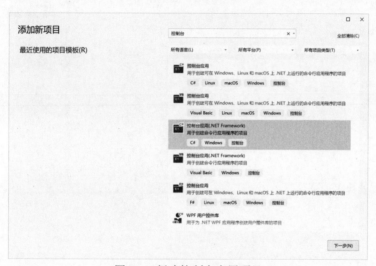

图 1-9 新建控制台应用项目

图 1-10 配置控制台应用项目

注意：

① 在输入项目的保存位置时，如果指定的文件夹不存在，Visual Studio 会自动创建。

② Visual Studio 2022 提供了控制台应用程序、Windows 窗体应用程序等各种类型应用程序的模板。控制台应用程序是指通过命令行运行的控制台应用，此种应用程序通过 DOS 环境下的命令行与用户进行交互。Windows 窗体应用程序将在第 4 章讲解。

3. 编写程序代码

控制台项目创建完成后会自动添加 Program 类，并自动生成一段程序代码，如图 1-11 所示。代码中可以看到 Main()方法，Main()方法是程序的入口，程序执行时从 Main()方法开始，可在 Main()方法中编写程序代码。

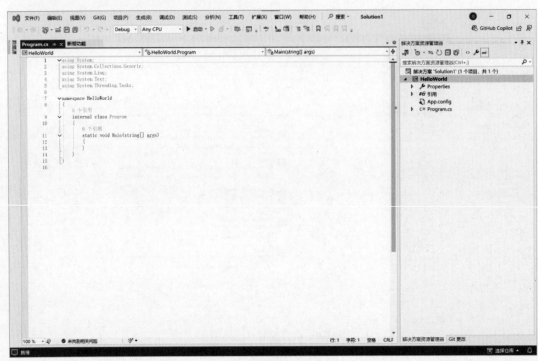

图 1-11 控制台应用项目

【示例代码：chapter01\Solution1\HelloWorld】

```
1  using System;
2  using System.Collections.Generic;
3  using System.Linq;
4  using System.Text;
5  using System.Threading.Tasks;
6  namespace HelloWorld
7  {
8     class Program
9     {
10        static void Main(string[] args
11        {
```

```
12           Console.WriteLine("HelloWorld");
                              //用于向控制台输出字符串"HelloWorld!"
13           Console.ReadLine();    //用于暂停程序
14        }
15    }
16 }
```

【分析】

（1）第1~5行：程序集中引用的命名空间。在编辑 C# 程序时，如果要使用.NET Framework 中的类，必须引入相应的命名空间。例如，在本实例中第 1 行的"using System;"表示引入 System 命名空间中的类。System 命名空间是.NET 最基本的命名空间，包含最基本的类的声明与实现。

（2）第6行：声明一个新命名空间，名称为 HelloWorld，新命名空间从第 7 行花括号开始，一直到第 16 行花括号结束。注意，花括号{}必须成对匹配，否则将出现编译错误。C#语言使用命名空间控制源程序代码的范围，以加强源程序代码的组织管理。例如，本行中的 namespace 是 C#关键字，用来声明命名空间，HelloWorld 是自定义的命名空间名称。Visual Studio 在创建应用程序项目时，自动使用项目名称设置命名空间的名称。

（3）第8行：关键字 class 用于声明一个类，Program 是类的名称，类的成员从第 9 行花括号开始，直到第 15 行花括号结束。C#是一个完全面向对象的语言，C#语言必须封装在类之中，一个程序至少包括一个自定义类。

（4）第10行：定义了一个 Main()方法，该方法是程序的入口，方法的成员从第 11 行花括号开始，一直到第 14 行花括号结束，该方法包含两行语句。C#控制台应用程序必须包含一个 Main()方法，在运行时，首先从 Main()方法的第 1 条语句开始执行，当最后一条语句被执行后，程序结束运行。本实例中的"Console.WriteLine("HelloWorld!");"就是一条语句，表示调用 System 命名空间中 Console 类的 WriteLine()方法，把字符串输出到控制台。Console 类包含了与控制台相关的输入输出方法，除 WriteLine()方法外，还有 ReadLine()等方法，ReadLine()方法表示从键盘缓冲区读取一行字符，本实例中使用 ReadLine()方法用于使程序暂停。

注意：

① 在编写程序代码时，要充分利用智能感知功能快速输入源程序代码，以避免输入错误。例如，在程序第 12 行中，在输入"Console."之后，系统将自动显示 Console 的所有成员列表，先滚动浏览该列表框或按 W 键，快速定位到 WriteLine，再按空格键，由系统自动完成 WriteLine 的选择和输入。

② C#语言严格区分大小写字母，因此输入代码时注意不要混淆大小写字母。

③ 在编写控制台应用程序时，一般会使用 Console.ReadLine()暂停程序，如果不编写此行代码，可以使用 Ctrl+F5 组合键，使程序不带调试直接运行，运行结束后控制台窗口不会关闭，按任意键后可以关闭。

④ 在编码时要注意代码的规范性和可读性，可以使用 Ctrl+K 组合键和 Ctrl+D 组合键，自动调整代码进行缩进。

4. 运行程序

选择"调试"→"启动调试"命令（或按 F5 键）可进行保存和运行，如果程序出现了错误，相应的编译结果及错误信息会分别显示在错误列表中。Shift+F5 组合键用于结束运行。对于控制台类型的应用程序来说，其基本的输入输出都是在一个命令行窗口中实现的。HelloWorld 程序运行结果如图 1-12 所示。

图 1-12　HelloWorld 程序运行结果

5. 排错

如果程序代码有错误，可在如图 1-13 所示的错误列表中双击错误信息，定位到可能出错的语句，排除错误后，重新运行即可。

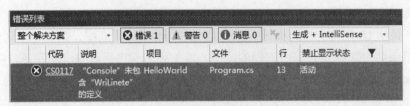

图 1-13　错误列表

1.3.2　使用 Visual Studio 的技巧

在使用 VS 时有很多小技巧，可以简便开发过程。下面简单介绍其中的一些技巧。

1. MSDN 帮助

MSDN（Microsoft Developer Network）是微软公司面向软件开发者提供的一种信息服务，包含联机帮助文件和技术文献。按 F1 键可以启动在线帮助程序，如图 1-14 所示。

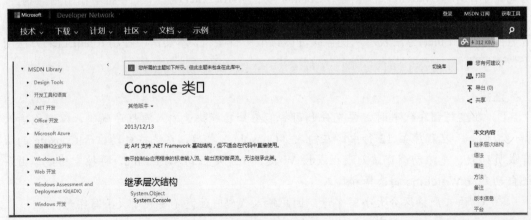

图 1-14　MSDN 帮助

2. 显示行号

选择"工具"→"选项"命令，将显示如图 1-15 所示的"选项"对话框，在"文本编辑器"→C#项中，选中"行号"复选框，可以在代码前面显示行号，便于调试和排错。

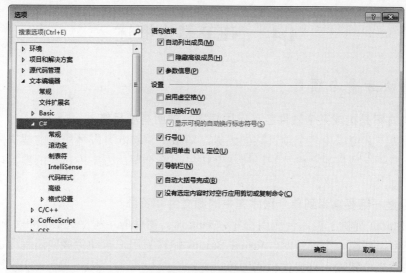

图 1-15　设置显示行号

3. 预置代码段

代码段是一类能够提高范式代码编写速度的代码，如果有些代码使用方式比较固定，为了避免每次大量的代码输入，可以将它们设置为代码段。

代码段的查看方法是：在需要插入代码段之处右击，在弹出的菜单中选择"插入代码段"命令，如图 1-16 所示，从中选择要插入的代码段即可。

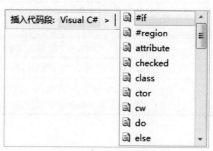

图 1-16　预置代码段

如果要插入已知存在的代码段，例如，for 循环代码段，可以直接在代码中输入 for，然后连续按两次 Tab 键即可完成代码段的输入。

4. 代码折叠

如果一个 .cs 文件中的代码比较多，当打开该文件时不方便快速定位，也不方便快速了解该文件所包含的功能。为了方便书写代码并能快速了解该文件情况，也为了后期维护方便，可以使用#region 和#endregion 实现代码折叠功能，从而实现类似目录或大纲视图的效果。其使用方式一般如下：

```
#region 说明文字
    //代码块
#endregion
```

1.4 .NET 基本概述

1.4.1 解决方案与项目

1. 解决方案是构成某个软件包（应用程序）的所有项目集

一个解决方案可以由多个项目共同组成。Visual Studio 开发环境中的解决方案资源管理器就是管理所有项目的文件，它以树状结构显示整个解决方案中包含的项目以及每个项目的组成信息。

2. 项目是一组要编译到单个程序集中的源文件和资源

Visual Studio 提供了控制台应用程序、Windows 窗体应用程序、ASP.NET Web 应用程序等项目类型，新建一个项目时，Visual Studio 能够为用户自动生成应用程序的框架，用户只需要在适当位置输入自己的代码即可。

表 1-2 列出了 Visual Studio 中的一些常用的文件类型。

表 1-2　Visual Studio 中常用的文件类型

扩展名	名称	描述
.sln	Visual Studio 解决方案	.sln 文件为解决方案资源管理器提供显示管理文件的图形接口所需的信息。打开.sln 文件，能快捷地打开整个项目的所有文件
.csproj	项目文件	一个特殊的 XML 文件，主要用来控制项目的生成
.cs	源代码文件	表示源代码文件、Windows 窗体文件、Windows 用户控制文件、类文件、接口文件等
.resx	资源文件	包括一个 Windows 窗体、Web 窗体等文件的资源信息
.aspx	Web 窗体文件	表示 Web 窗体，由 HTML 标记、Web Server 控件及脚本组成
.asmx	XML Web 服务文件	表示 Web 服务，它链接一个特定.cs 文件，在这个.cs 文件中包含了供 Internet 调用的方法代码

1.4.2 命名空间

为了便于组织和管理，C#语言引入了命名空间的概念。命名空间相当于一个容器，包含一组定义的类或结构。命名空间也可以嵌套在另一个命名空间中，具有相同名称的类位于不同的命名空间。.NET 类库中常用的命名空间如表 1-3 所示。

要调用类，可以使用以下两种方法。

（1）通过命名空间直接调用。

命名空间.类名.方法名(参数)

具体示例：

System.Console.WriteLine("HelloWorld!");

该语句调用了 System 命名空间下 Console 类的静态方法 WriteLine，用来在控制台输出字符串 "HelloWorld!"。

表 1-3 .NET 类库中常用的命名空间

命 名 空 间	类 的 功 能
System	包含基本类和基类，定义常用的数据类型、数据转换、数学运算、异常处理等
System.Data	主要包括了组成 ADO.NET 体系结构的类
System.IO	包含允许读写文件和数据流的类型以及提供基本文件和目录支持的类型
System.Drawing	提供对 GDI+基本图形功能的访问
System.Net	为网络上使用的多种协议提供简单的编程接口
System.Threading	提供一些使得可以进行多线程编程的类和接口
System.Windows.Forms	包含创建基于 Windows 的应用程序的类，以利用操作系统中提供的界面功能
System.Web	提供使得可以进行浏览器与服务器通信的类和接口
System.Xml	提供对 XML 文件进行处理的类

（2）先引入命名空间后再调用。

using 命名空间;

具体示例：

```
using System;
Console.WriteLine("HelloWorld!");
```

在程序的开头使用关键字 using 引入命名空间，然后在需要时直接调用该命名空间下的 Console 类即可。

注意：如果按照第二种方法，不同的命名空间下有相同名称的类要使用，则编译器会提示错误，此时必须按照第一种方法添加命名空间来防止二义性的出现。

1.4.3 Main()方法

一个程序只能有一个 Main()方法入口。程序的功能是通过执行方法的代码实现的，每个方法都是从第一行开始至最后一行，中间可以调用其他方法，以完成各种各样的操作。每个应用程序都可能包含了多个方法，但入口方法只有一个。每个控制台程序和 Windows 应用程序中，必须有一个类包含名为 Main 的静态方法。例如：

```
static void Main(string[] args)
{
}
```

string[] 是声明一个字符串数组类型，args 就是这个数组的变量名称，也是该方法的参数。

注意：

① Main()方法首字母必须大写。

② Main()方法是可执行程序的入口点，并且入口点是唯一的。

③ 程序在 Main()方法中开始，也在 Main()方法中结束，对应的线程为主线程。

④ Main()方法一般都是 void 类型的，但也可以声明为 int 类型。

⑤ Main()方法可以带一个字符串数组参数，也可以不带参数。

⑥ Main()方法可以声明为 static，也可以声明为非 static 类型。

1.4.4 代码注释

为源代码加上注释,不但可以提高程序的可读性,而且便于后期的维护。C#语言中的注释方法有以下 3 种。

1. 单行注释

单行注释通常用于对程序中的某一行代码进行解释,用符号"//"表示,后面为被注释的内容。具体示例如下。

```
Console.WriteLine("HelloWorld!");    //用于向控制台输出字符串"HelloWorld!"
```

2. 多行注释

多行注释就是注释中的内容可以为多行,它以符号"/*"开头,以符号"*/"结尾。具体示例如下。

```
/* 用于向控制台输出字符串"HelloWorld!"
   用于暂停程序  */
```

3. 文档注释

文档注释用于对类或方法进行说明和描述。在类或方法上一行连续输入 3 个"/",就会自动生成相应的文档注释,用户需要手动填写类或方法的描述信息,完成文档注释的内容,具体示例如下。

```
/// <summary>
/// 控制台输出字符串"HelloWorld!"
/// </summary>
```

1.4.5 程序调试

在程序开发过程中,可能会出现各种各样的错误,Visual Studio 自带了调试功能,用于快速定位错误信息。

1. 单步调试

在程序出现错误时,可使用单步调试来一步一步跟踪程序执行的流程,根据变量的值找出错误原因。单步调试分为两种,分别是逐语句调试和逐过程调试。逐语句调试使用 F11 键,进入到方法内部,单步执行方法体中的每条语句;逐过程调试使用 F10 键,不进入方法体内部,将方法作为一条语句来执行。

下面按 F11 键启动 HelloWorld 程序的单步调试,调试界面如图 1-17 所示。

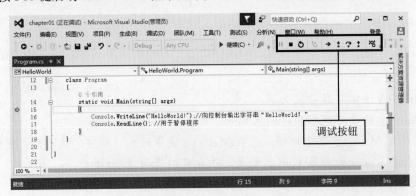

图 1-17 单步调试

从图 1-17 可以看出，当程序开启单步调试时会暂停到断点处，并出现一个箭头指向程序执行的位置，这时可以通过工具栏上的调试按钮执行相应的调试操作。图 1-17 中所标注的调试按钮功能如下。

（1）全部中断：用于将正在执行的程序全部中断。

（2）停止调试：用于停止调试程序。

（3）重新启动：用于重新启动程序调试。

（4）显示下一语句：用于显示下一条执行的语句。

（5）逐语句：用于让程序按照逐语句进行调试。

（6）逐过程：用于让程序按照逐过程进行调试。

（7）跳出：用于跳出正在执行的程序。

2. 设置断点

在程序调试过程中，断点可以使正在运行的程序在断点处暂停，可以通过设置断点来观察某些数据的变化情况，以便发现程序出错的原因。设置断点有以下两种方法。

（1）单击代码左边的灰色区域，断点插入成功后左侧会出现红色圆点，如图 1-18 所示。插入断点后的代码会高亮显示。相同操作可以删除断点。

图 1-18　插入断点方法 1

（2）在需要插入断点的代码处右击，在弹出的菜单中选择"断点"→"插入断点"命令，如图 1-19 所示。

图 1-19　插入断点方法 2

3. 条件断点

条件断点可以在设置断点的基础上，快速定位到需要调试的循环次数，下面使用 for 循环案例进行演示。

【示例代码：chapter01\Solution1\ForProgram】

```
namespace ForProgram
{
    class Program
    {
        static void Main(string[] args)
        {
            for (int i = 1; i <= 100; i++)
            {
                Console.WriteLine(i);
            }
            Console.ReadLine();
        }
    }
}
```

首先在需要中断的代码行前面添加断点，然后在断点上右击"条件"，在"断点条件"窗口中输入中断表达式"i==96"，如图 1-20 所示。

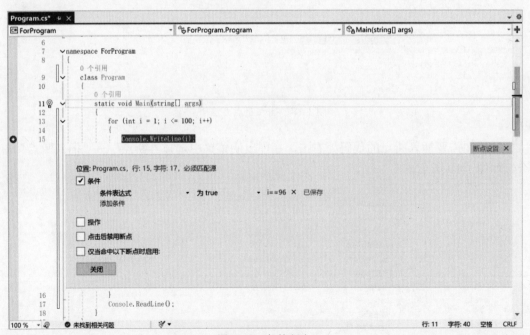

图 1-20　条件断点

启动调试，程序运行结果如图 1-21 所示。当循环中的变量条件符合条件断点中设置的条件时，循环将中断。

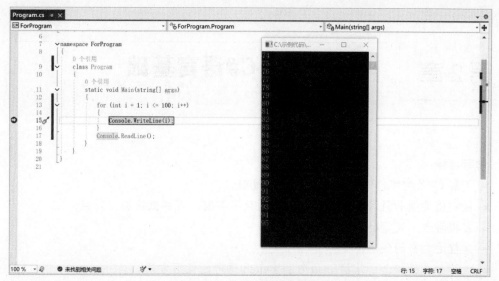

图 1-21　条件断点程序调试

4. 观察变量

在程序运行出错时，如果想尽快找到程序出错的原因，最简单的方法是观察当前变量的值，下面介绍两种观察变量值的方法。

（1）使用局部变量窗口。在调试过程中，在菜单栏中选择"调试"→"窗口"→"局部变量"命令，即可打开"局部变量"对话框查看变量的值，如图 1-22 所示。

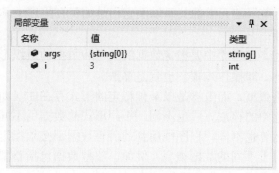

图 1-22　局部变量窗口

（2）使用鼠标悬停。在调试过程中，当需要查看变量的当前值时，将鼠标移动到当前变量所在的位置即可。

习题1

第 2 章　C#语言基础

学习目标：
- 了解 C#关键字，掌握标识符的命名规则；
- 区别值类型和引用类型，掌握常用的数据类型，理解数据类型转换；
- 掌握常量、变量的声明和应用；
- 掌握运算符的使用；
- 理解分支的概念，掌握 if 语句和 switch 语句的使用方法；
- 理解循环的概念，掌握循环语句的使用方法；
- 掌握跳转语句的使用方法；
- 掌握异常处理的方法。

2.1　C#语法元素

C#的基本语法包括数据类型、常量和变量、运算符和表达式、流程控制语句及异常处理等内容。下面分析这些语法元素的关系以及它们是如何在程序中体现的。一个程序应该包含以下两方面的内容：

（1）对数据的描述。在程序中要指定数据的类型和数据的组织形式，即数据结构。

（2）对操作的描述。即操作步骤，也就是算法。

算法处理的对象是数据，而数据是以某种特定的形式存在的（如整数、字符等形式），不同类型的数据在内存中的存放方式也不同。不同形式的数据用不同的数据类型表示，不同的数据类型参与的运算也不同。程序中用到的所有数据都必须指定其数据类型，有常量和变量之分，它们都属于不同的数据类型。例如，整型数据包括整型常量和整型变量。用运算符将运算对象连接起来，且符合 C#语法规则的式子称为表达式。算法是通过程序中的操作语句体现的。异常处理是当程序执行时遇到错误或意外情况时的处理机制，它能捕获异常，以便程序员定位错误，改正程序。

2.2　关键字与标识符

关键字是对 C# 编译器具有特殊意义的预定义保留标识符。标识符用来为程序中各元素定义名字。

2.2.1　关键字

C#中的关键字如表 2-1 所示，每个关键字都有特殊的作用。例如，using 关键字用于引入命名空间，class 关键字用于声明一个类。

注意：
① 所有关键字都是小写的。
② 程序中的标识符不能以关键字命名，除非加上@前缀。

表 2-1 C#关键字

abstract	const	extern	int	out	short	typeof
as	continue	false	interface	override	sizeof	uint
base	decimal	finally	internal	params	stackalloc	ulong
bool	default	fixed	is	private	static	unchecked
break	delegate	float	lock	protected	string	unsafe
byte	do	for	long	public	struct	ushort
case	double	foreach	namespace	readonly	switch	using
catch	else	goto	new	ref	this	virtual
char	enum	if	null	return	throw	volatile
checked	event	implicit	object	sbyte	true	void
class	explicit	in	operator	sealed	try	while

2.2.2 标识符

标识符是编程人员为常量、变量、方法、类等定义的名字。标识符最多可以由 511 个字符组成，这些字符可以是任意顺序的大小写字母、数字、下画线"_"和@符号，但不能以数字开头，且不能与关键字同名。

下面是合法的标识符：

```
username
user_pwd
_user12
@class
```

下面是不合法的标识符：

```
user!
0_user
user 12
class
```

注意：为了增加代码的可读性，建议编程人员在定义标识符时应该遵循以下规范。
（1）使用 Pascal 命名法定义类名、方法名和属性名，即每个单词的首字母要大写，如 Age、UserName、ArrayList。
（2）使用 Camel 命名法定义字段名和变量名，即首字母小写，之后的每个单词的首字母均为大写，字段名前加下画线"_"，如变量 arrayList、字段 _userName。
（3）常量名中的所有字母都大写，且单词之间使用下画线连接，如 USER_AGE。

（4）在定义标识符时，使用代表某种意义的英文单词，增加可读性。例如，使用 userName 表示用户名，使用 password 表示密码。

2.3 数据类型

数据类型是表示具有多种相同特征的一组数据。C#是强类型语言，即每个变量和对象都必须声明数据类型，并且为变量赋值时必须赋予与变量同一类型的值，否则程序会报错。

从用户的角度，数据类型分为内置数据类型和用户自定义的数据类型。内置数据类型是.NET Framework 预定义好的类型；用户自定义类型是由用户声明创建的。

从数据存储的角度，数据类型又分为值类型和引用类型。值类型用于存储数据的值，而引用类型用于存储对实际数据的引用地址。这两类又细分为多种数据类型，具体如表 2-2 所示。

表 2-2 C#数据类型

类型		说明
值类型	简单值类型	有符号整型：sbyte、short、int、long
		无符号整型：byte、ushort、uint、ulong
		浮点型：float、double
		高精度小数：decimal
		布尔型：bool
		Unicode 字符：char
	结构	struct S{…}形式的用户自定义类型
	枚举	enum E{…}形式的用户自定义类型
引用类型	Object	所有其他类型的最终基类型：object
	类	class C{…}形式的用户自定义类型
	接口	interface I{…}形式的用户自定义类型
	字符串	Unicode 字符串：string
	数组	一维和多维数组
	委托	delegate TD{…}形式的用户自定义类型

值类型与引用类型的不同之处在于：值类型的变量直接包含数据；而引用类型的变量存储对数据的引用，后者称为对象。对于值类型，每个变量都有自己的数据副本，各变量之间的操作互不影响；对于引用类型，同一个对象可能被多个变量引用，因此一个变量的操作可能影响另一个变量对该对象的引用。在定义一个值类型变量后，将直接为该类型分配空间，可以直接赋值和使用；而引用类型在定义时并不会分配空间，只有在对其实例化时，才真正地分配存储空间。

注意：在 C#中，若没有为结构或类的字段变量初始化，则编译器会自动根据这些数据类型为其赋一个默认值。默认值规则如下：

① 数值（整数和小数）：0。
② 字符：'\0'对应的代码点为 U+0000。

③ 布尔：false。

④ 枚举：0。

⑤ 引用类型：null。

2.3.1 简单值类型

C#语言提供了一组已经定义的简单值类型，它们都是具有唯一取值的数据类型，从计算机表示的角度可分为数值类型、布尔类型和字符类型。

1. 数值类型

根据不同的范围、精度和用途，C#语言将数值类型分为整数型、浮点型和小数型，具体描述如表 2-3 所示。其中，别名是指该类型对应的.NET Framework 中数据类型的全名。

表 2-3　C#中的数值类型

类型	名　称	别　　名	取 值 范 围	空间	说　　明
整数型	sbyte	System.Sbyte	$-128\sim127$	1B	有符号 8 位整数
	byte	System.Byte	$0\sim255$	1B	无符号 8 位整数
	short	System.Int16	$-32\,768\sim32\,767$	2B	有符号 16 位整数
	ushort	System.UInt16	$0\sim65\,535$	2B	无符号 16 位整数
	int	System.Int32	$-2\,147\,483\,648\sim2\,147\,483\,647$	4B	有符号 32 位整数
	uint	System.UInt32	$0\sim42\,994\,967\,295$	4B	无符号 32 位整数
	long	System.Int64	$-9\,223\,372\,036\,854\,775\,808\sim9\,223\,372\,036\,854\,775\,807$	8B	有符号 64 位整数
	ulong	System.UInt64	$0\sim18\,446\,744\,073\,709\,551\,615$	8B	无符号 64 位整数
浮点型	float	System.Single	$\pm1.5\times10^{-45}\sim\pm3.4\times10^{38}$	4B	32 位单精度实数
	double	System.Double	$\pm5.0\times10^{-324}\sim\pm1.7\times10^{308}$	8B	64 位双精度实数
小数型	decimal	System.Decimal	$\pm1.0\times10^{-28}\sim\pm7.9\times10^{28}$	16B	128 位十进制实数

（1）整数型变量用来存储整数数值，也就是说，整数型的数据值只能是整数。例如，1 为一个整数，而 1.0 则不是一个整数。数学上的数可以是负无穷大到正无穷大，但由于计算机内存的存储单元有限，C#所提供的数据都是有一定范围的，它们的取值范围见表 2-3。

给整数型的变量赋值时，可采用十进制或十六进制的数值常数，十六进制的数值需加前缀 0x。要定义一个变量为 long 类型，需要在数值后面加上类型指示符 L（或 l）。

例如：

```
int x = 123;
long y = 0x12aL;
```

（2）浮点型变量用来存储小数数值，双精度符点数 double 所表示的符点数比单精度符点数 float 更精确。给浮点型的变量赋值时，默认情况下赋值运算符右侧的数值被视为 double 型，如果使用 float 型，需要在数值后面加上类型指示符 F（或 f）。

例如：

```
float x = 3.14F;
```

```
double y = 4.5;
```

（3）小数型 decimal 是适合财务和货币类型计算的 128 位的数据类型。要定义一个变量为 decimal 类型，需要在数值后面加上类型指示符 M（或 m）。

例如：

```
decimal myNum = 10.45M;
```

注意：定义变量时要选择合适的数据类型，过长的类型会浪费存储资源，过短的类型不足以满足变量的变化范围。在整数类型中，尤其对有符号类型，应优先考虑使用 int 型，int 型消耗内存少，运行速度快。

2. 布尔类型

布尔类型（bool）只有 true（真）和 false（假）两个值，不能用数据类型中的 1 或 0 代替，在计算机中占 4B，主要应用到数据运算的流程控制中，辅助实现逻辑分析和推理。

例如：

```
bool i = true;
```

注意：在 C#中，布尔类型值只有两个，true 和 false。这与 C 和 C++不同，在 C 和 C++中，0 表示"假"，任何非 0 值都表示"真"。因此，C#中的布尔类型与整型是完全不同的。例如，"bool bl=0;"这样定义布尔类型变量是错误的。

3. 字符类型

字符类型（char）即保存单个字符的值，每个字符对象与 Unicode 字符集中的字符相对应。char 类型的变量值需要用单引号括起来。

例如：

```
char c1 = 'A';
```

除了以上形式的字符外，C#还允许使用一些特殊形式的字符常量，即以一个字符"\"开头的字符序列。这种特殊的字符序列称为转义字符，用来在程序中取代特殊的控制字符。转义字符在屏幕上是不能显示的，并且也无法用一般形式的字符表示。C#定义的转义字符如表 2-4 所示。

表 2-4　转义字符

转义字符	含义	转义字符	含义
\'	单引号	\r	回车，将当前位置移动到本行开头
\"	双引号	\b	退格，将当前位置移到前一列
\\	反斜杠	\f	换页，将当前位置移动到下一页开头
\0	空字符	\n	换行，将当前位置移动到下一行开头
\a	感叹号	\x	后面跟 2 个十六进制数字，表示一个 ASCII 字符
\t	水平制表符	\u	后面跟 4 个十六进制数字，表示一个 Unicode 字符

2.3.2 结构

上述几种简单值类型可以完成基本的数据运算。只有这些基本数据类型是不够的,有时要将不同类型的数据组合成一个整体,如一名学生的学号、姓名、成绩等。C#提供了结构类型(struct)解决这个问题。

结构通常用来封装小型相关变量组,如学生的成绩、商品的类型等。在C#中,作为一个整体的学生(如名称为 Student)称为结构,而学生的姓名、学号、成绩等数据项称为结构的成员。

结构类型是使用关键字 struct 定义的,声明结构类型的语法格式如下:

```
[访问修饰符] struct 结构名称
{
    结构体;
}
```

【例 2-1】 声明一个结构 Student,其中 name、id 和 score 是结构的数据成员,并在 Main()方法中使用该结构赋值并输出。

【示例代码:chapter02\Solution1\StructType】

程序代码如下:

```
namespace StructType
{
    //定义测试类
    class Program
    {
        static void Main(string[] args)
        {
            Student s;                          //使用结构类型
            s.name = "张三";
            s.id = 2024800001;
            s.score = 93;
            Console.WriteLine("{0}的学号是{1}, .NET 程序设计课程成绩是{2}分。",
                s.name, s.id, s.score);
            Console.Read();
        }
    }
    //定义结构 Student
    struct Student
    {
        public string name;
        public int id;
        public int score;
    }
}
```

程序运行结果如图 2-1 所示。

图 2-1 StructType 项目运行结果

【分析】

本实例中定义了一个结构类型，类型名称为 Student，它包含 3 个成员，其中 public 为各成员的访问权限。为了在程序中使用该结构类型的数据，定义了 Student 类型的变量 s，并分别为变量 s 的 name、id 和 score 赋值并输出。

程序在输出时采用了一种新的方式，即占位符的方式。{0}、{1}、{2}即是占位符，它们位置会被后续的变量列表依次替代。例如，此例中，{0}将被变量 s 的 name 值替代，{1}将被变量 s 的 id 值替代，{2}将被变量 s 的 score 值替代。

注意：C#内置的结构类型主要有 DateTime 和 TimeSpan 等。DateTime 表示某个时间点，其成员主要有 Year、Month、Day、Hour、Minute、Second、Today、Now 等，分别表示年、月、日、时、分、秒、今天、当前时间。TimeSpan 表示某个时间段，其成员主要有 Days、Hours、Minutes、Seconds 等，分别表示某个时间段的天数、小时数、分数、秒数。

2.3.3 枚举

枚举（enum）是一种由一组被称为枚举数列表的常数所组成的独特类型。每种枚举类型都有对应的数据类型，可以是除 char 以外的任何简单数据类型。

枚举类型是使用关键字 enum 定义的，声明枚举类型的语法格式如下：

```
[访问修饰符] enum 枚举类型名称
{
    枚举列表
};
```

例如，当数字 0、1、2、3、4、5、6 表示星期时，为直观起见，可以先使用一组中文符号表示，依次为星期日、星期一、星期二、星期三、星期四、星期五、星期六，并给它们定义一个统一的名称如 Weekdays，使用 enum 标记，完整程序代码如下：

```
enum Weekdays {星期日, 星期一, 星期二, 星期三, 星期四, 星期五, 星期六};
```

其中，Weekdays 是枚举型的名称，而花括号中的中文字符分别表示 7 个不同的枚举元素。

【**例 2-2**】 声明一个枚举类型 Weekdays（枚举列表中成员的默认类型为 int，其中星期日是第一个枚举数，值为 0，后面每个枚举数的值依次递增 1），使用该枚举类型并输出。

【示例代码：chapter02\Solution1\EnumType】

程序代码如下：

```
namespace EnumType
{
    //定义测试类
    class Program
    {
        static void Main(string[] args)
        {
            Weekdays myday = (Weekdays)DateTime.Now.DayOfWeek;//使用枚举类型
            Console.WriteLine("今天是：{0}",myday);
            Console.ReadLine();
        }
    }
    //定义枚举类型Weekdays
    enum Weekdays {星期日,星期一,星期二,星期三,星期四,星期五,星期六};
}
```

程序运行结果如图 2-2 所示。

【分析】

本实例中定义了枚举类型,枚举名为 Weekdays,枚举值一共 7 个,即一周中的 7 天,所有被声明为 Weekdays 类型的变量值只能是 7 天中的某一天。

图 2-2　EnumType 项目运行结果

注意：

① 枚举元素的数据值是确定的,一旦声明,就不能在程序的运行过程中更改。

② 枚举元素的个数是有限的,同样一旦声明,就不能在程序的运行过程中增加或减少。

③ 默认情况下,枚举的值是一个整数,第一个枚举数的值默认为 0,也可以自定义更改,后面每个枚举数的值依次递增 1。

④ 如果需要修改默认的规则,则重写枚举元素的值即可。例如：

```
enum EnumType {a=7,b,c=12,d,e};
```

在此枚举型中,a 的值为 7,b 为 8,c 为 12,d 为 13,e 为 14。

⑤ 枚举型与结构型是有区别的。结构实质上是若干数据成员与数据操作的组合,一个结构型的值是由各个成员的值组合而成的,结构型各个数据成员的数据类型可以是不同的,如在例 2-1 中的结构型变量 s 的值是由"张三"、2024800001 与 93 这 3 个数据构成的。枚举型的各个枚举元素的数据类型是相同的,枚举数只能代表某一个枚举元素的值,如在例 2-2 中的枚举变量 myday 在程序中只代表枚举元素星期四,其值为 4。

2.3.4 Object

Object 是 C# 语言中所有数据类型的根类，即所有类型都是直接或间接从 Object 继承的。

【例 2-3】 定义 Object 类型的变量 a，并将不同类型的值赋予 a 并输出。

【示例代码：chapter02\Solution1\ObjectType】

程序代码如下：

```
namespace ObjectType
{
    class Program
    {
        static void Main(string[] args)
        {
            object a;        //定义一个Object类型的变量a
            a = 10;          //将整型10赋值给a
            Console.WriteLine(a);
            Console.WriteLine(a.GetType());
            a = false;       //将布尔型false赋值给a
            Console.WriteLine(a);
            Console.WriteLine(a.GetType());
            a = "ABC";       //将字符串型ABC赋值给a
            Console.WriteLine(a);
            Console.WriteLine(a.GetType());
            Console.ReadLine();
        }
    }
}
```

程序运行结果如图 2-3 所示。

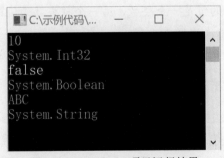

图 2-3 ObjectType 项目运行结果

【分析】

本实例中，分别为 Object 类型的变量 a 赋值，并调用 GetType() 方法输出当前 a 的数据类型。其中，10 的类型是 32 位整数型，false 为布尔类型，ABC 为字符串型。

注意：在 C#中有一些特殊的转换，其转换发生在值类型和 Object 类型之间，这就是封箱与拆箱。所谓封箱，是指将值类型转换为 Object 类型，封箱是隐式进行的；拆箱，则是指将 Object 类型转换为值类型，拆箱是显式进行的。频繁的装箱和拆箱会影响程序的性能，因此要慎重使用。

2.3.5 类

类是能存储数据并执行操作的复杂数据类型，用来定义对象的可执行操作（如属性和方法等），类的实例是对象。

类是使用关键字 class 定义的，声明类的语法格式如下：

```
[访问修饰符] class 类名称
{
    类成员;
}
```

【例 2-4】 声明一个基类 Student，声明一个派生类 ComputerStudent，ComputerStudent 继承了 Student 类的两个字段_name 与_id，而又有自己独有的一个字段_computerScore。

【示例代码：chapter02\Solution1\ClassType】

程序代码如下：

```
namespace ClassType
{
    //定义测试类
    class Program
    {
        static void Main(string[] args)
        {
            ComputerStudent s = new ComputerStudent();
            s._name = "张三";
            s._computerScore = 93;
            Console.WriteLine("{0}的计算机课程成绩是{1}", s._name,
            s._computerScore);
            Console.Read();
        }
    }
    //定义父类Student
    class Student
    {
        public string _name;
        public int _id;
    }
    //定义子类ComputerStudent
    class ComputerStudent: Student
    {
```

```
        public int _computerScore;
    }
}
```

程序运行结果如图 2-4 所示。

图 2-4　ClassType 项目运行结果

【分析】

本实例中,分别定义了父类 Student 与子类 ComputerStudent,其中子类 ComputerStudent 继承了父类的姓名字段和学号字段。在访问类成员时,首先对 ComputerStudent 类进行了实例化,通过对象 s 为类成员赋值和输出。

注意:

① 类名是一个合法的 C#标识符,推荐使用 Pascal 命名规范,Pascal 命名规范要求名称的每个单词的首字母大写。

② 在访问类成员时,一定要先对对象进行实例化。如果未对对象 ComputerStudent 进行实例化,而直接访问其成员,编译时将出现"使用了未赋值的局部变量 's'"的错误。

③ 类是抽象的,不占用内存,而对象是具体的,占用存储空间。

2.3.6　接口

接口(interface)是指描述可属于任何类或结构的一组相关功能。接口可由属性、方法、事件、索引器任意组合构成,但接口不能包含字段。接口不能单独存在,不能实例化,只能在实现接口的类中实现。

接口与类的区别:比喻来说,接口类似做指示的高层领导,而类是一线工作人员。因为接口从不做具体的实现,只是指导性的东西(如方法的声明等,即定义基本的功能框架)。而类一旦接受某个接口的领导(即实现某个接口),那么就得遵循指导性方案,把领导说过的每一件事办妥(即实现接口中的方法、属性、索引等,因为在接口中只能大概声明,没有具体实现,等着一线工作人员完成)。

接口是使用关键字 interface 定义的,声明接口的语法格式如下:

```
[访问修饰符] interface 接口名称
{
    接口成员;
}
```

【例 2-5】 声明一个接口 Animal,在接口中定义了一个抽象方法 Bark(),并定义一个类 Dog 实现接口中的所有方法。

【示例代码：chapter02\Solution1\InterfaceType】
程序代码如下：

```csharp
namespace InterfaceType
{
    //定义测试类
    class Program
    {
        static void Main(string[] args)
        {
            Dog dog = new Dog();
            dog.Bark();
            dog.Run();
            Console.ReadLine();
        }
    }
    //定义Animal接口
    interface Animal
    {
        void Bark();
    }
    //定义Dog类实现Animal接口
    class Dog : Animal
    {
        public void Bark()
        {
            Console.WriteLine("狗在叫！");
        }
        public void Run()
        {
            Console.WriteLine("狗在跑！");
        }
    }
}
```

程序运行结果如图 2-5 所示。

图 2-5　InterfaceType 项目运行结果

【分析】

本实例中定义了 Animal 接口，并定义 Dog 类继承 Animal 接口，Dog 类的 Bark() 方法实现了 Animal 接口的 Bark() 方法。同时，Dog 类中也定义了自己的方法 Run()。

注意：对于接口中定义的成员有如下要求。

① 接口的成员必须是方法、属性、事件或索引器。接口不能包含常量、字段、运算符、构造方法以及任何类的静态成员等。

② 接口不提供对它所定义成员的实现，实现由继承的类完成。

③ 接口成员都是 public 类型的，但不能使用 public 修饰符。

④ 一个类虽然只能继承一个基类，但可以实现任意数量的接口。

2.3.7 字符串

字符串是一个由若干个 Unicode 编码组成的字符数组。字符串由关键字 string 声明，字符序列使用一对双引号引起来，如".NET 程序设计""HelloWorld"等都是字符串。

例如：

```
string a = "string1";
```

以上代码表示声明了一个字符串变量 a。

两个字符串可能通过加号运算符连接。

例如：

```
string a = "string1";
string b = "a="+a;
```

以上代码中，字符串 b 的值为"a=string1"，就是将两个字符串"a="和"string1"连接起来了。

字符串可以看成一个字符数组，允许通过索引提取字符串中的字符。

例如：

```
string a = "string1";
char x = a[0];
```

以上代码执行后，字符型变量 x 的值为 's'。

允许使用关系运算符==、!=比较两个字符串所对应的字符是否相同。

例如：

```
string a = "string1";
string b = "string2";
```

则 a==b 的运算结果为 false。

【例 2-6】 字符串赋值与字符串连接。

【示例代码：chapter02\Solution1\StringType】

程序代码如下：

```
namespace StringType
{
    class Program
    {
        static void Main(string[] args)
        {
```

```
            string a = "string1";
            string b = "string2";
            Console.WriteLine("a = "+a);
            Console.WriteLine("b = "+b);
            Console.WriteLine(a + b);
            Console.ReadLine();
        }
    }
}
```

程序运行结果如图 2-6 所示。

图 2-6　StringType 项目运行结果

【分析】

本实例中定义了两个字符串变量 a 和 b，并使用加号进行字符串拼接，分别输出 a 的值、b 的值以及 a 和 b 拼接后的值。

注意：C#的 string 是 System.String 的别名。在.NET Framework 中，System.String 提供的常用属性和方法有获得字符串长度的 Length、复制字符串的 Copy、从左查找字符的 IndexOf、从右查找字符的 LastIndexOf、插入字符的 Insert、删除字符的 Remove、替换字符的 Replace、分割字符的 Split、取子字符串的 Substring、压缩字符串空白的 Trim 以及格式化字符串的 Format 等。

2.3.8　数组

数组是由若干数据类型相同的数组元素构成的数据结构，索引从 0 开始，每个元素可以通过数组名称和索引进行访问。数组元素可以是任何类型，但因为没有名称，只能通过索引（又称下标）来访问。数组有一个"秩"，它表示和每个数组元素关联的索引个数。数组的秩又称为数组的维度。"秩"为 1 的数组称为一维数组，"秩"大于 1 的数组称为多维数组，根据维度，多维数组分为二维数组、三维数组等。常用声明数组的语法如表 2-5 所示。

表 2-5　常用声明数组的语法

数 组 类 型	声 明 语 法
一维数组	数据类型[] 数组名
二维数组	数据类型[,] 数组名
三维数组	数据类型[,,] 数组名

下面分别介绍一维数组和多维数组的创建和初始化。

1. 一维数组

（1）一维数组的声明与创建。

声明和创建一维数组的一般形式如下：

数组类型[] 数组名 = new 数组类型[数组长度];

例如：

int[] arr1 = new int[5];

表示声明和创建一个具有 5 个元素的一维数组 arr1。一维数组也可以先声明后创建。

（2）一维数组的初始化。

如果在声明和创建数组时，没有初始化数组，则数组元素将自动初始化为该数组类型的默认初始值。初始化数组有以下 3 种方式。

① 创建时初始化。

在创建一维数组时，对其初始化的一般形式如下：

数组类型[] 数组名 = new 数组类型[数组长度]{初始值列表};

其中，数组长度可省略。如果省略数组长度，系统将根据初始值的个数确定一维数组的长度。如果指定了数组长度，则 C#要求初始值的个数必须和数组长度相同，也就是所有数组元素都要初始化，而不允许只对部分元素进行初始化。初始值之间用逗号间隔。

例如：

int[] arr1 = new int[5] { 1, 2, 3, 4, 5 };

以上代码表示创建一个一维数组 arr1 具有 5 个数组元素，也可以在创建一维数组及初始化时采用简写形式。

例如：

int[] arr1 = { 1, 2, 3, 4, 5 };

以上代码同样表示创建了数组元素值分别为 1、2、3、4、5 的一个具有 5 个数组元素的一维数组。

② 先声明后初始化。

C#允许先声明一维数组，然后再初始化各数组元素。其一般形式如下：

数组类型[] 数组名;
数组名 = new 数组类型[数组长度]{初始值列表};

例如：

int[] arr1;
arr1 = new int[5] { 1, 2, 3, 4, 5 };

以上代码表示先声明一个一维数组 arr1，再用关键字 new 创建并初始化。注意，在先声明后初始化数组时，不能采用简写形式。

例如：

```
int[] arr1;
arr1 = { 1, 2, 3, 4, 5 };
```

代码是错误的。

③ 先创建后初始化。

C#也允许先声明和创建一维数组，然后逐个初始化数组元素。其一般形式如下：

数组类型[] 数组名 = new 数组类型[数组长度];
数组元素 = 值;

例如：

```
int[] arr1 = new int[5];
arr1[0] = 1;
arr1[0] = 2;
```

2. 多维数组

（1）多维数组的声明和创建。

声明和创建多维数组的一般形式如下：

数组类型[逗号列表] 数组名 = new 数组类型[数组长度];

其中，逗号列表的逗号个数加1就是维度数，即如果逗号列表为一个逗号，则称为二维数组；如果为两个逗号，则称为三维数组，依次类推。维度长度列表中的每个数字定义维度的长度，数字之间以逗号间隔。

例如：

```
int[,] arr2 = new int[3, 2];
```

以上代码表示声明和创建一个具有3×2共6个数组元素的二维数组arr2。

（2）多维数组的初始化。

多维数组也具有多种初始化方式，但需要注意以下几点：

① 以维度为单位组织初始化值，同一维度的初始值放在一对花括号中{}中。

例如：

```
int[,] arr2 = new int[3, 2] { { 1, 2 }, { 3, 4 }, { 5, 6 } };
```

② 可以省略维度长度列表，系统能够自动计算维度和维度的长度，但逗号不能省略。

例如：

```
int[,] arr2 = new int[, ] { { 1, 2 }, { 3, 4 }, { 5, 6 } };
```

③ 多维数组不允许部分初始化。

例如：

```
int[,] arr2 = new int[3, 2] { { 1, 2 }, { 3, 4 } };
```

以上代码希望只初始化二维数组的前两行元素，这是错误的。

【例 2-7】 声明一个一维数组和一个二维数组，并输出指定元素的值。

【示例代码：chapter02\ Solution1\ArrayType】

程序代码如下：

```csharp
namespace ArrayType
{
    class Program
    {
        static void Main(string[] args)
        {
            //声明一维数组
            int[] arr1;
            //创建数组对象
            arr1 = new int[5] { 1, 2, 3, 4, 5 };
            //声明二维数组并创建数组对象
            int[,] arr2 = new int[3, 2] { { 1, 2 }, { 3, 4 }, { 5, 6 } };
            //输出指定元素的值
            Console.WriteLine("arr1[0]的值是:"+ arr1[0]);
            Console.WriteLine("arr2[0, 1]的值是:"+arr2[0, 1]);
            Console.ReadLine();
        }
    }
}
```

程序运行结果如图 2-7 所示。

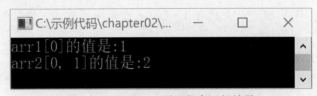

图 2-7 ArrayType 项目程序运行结果

【分析】

本实例中声明了一维数组 arr1 和二维数组 arr2，分别为数组 arr1 和 arr2 赋值，并使用数组名[索引]的形式获得数组中元素值，其中 arr1[0]指 arr1 的第一个值 1，arr2[0, 1]指 arr2 中第一行第二列的值 2。

注意：C#的数组类型是从抽象基类型 System.Array 派生的。Array 类的 Length 属性返回数组长度，Array 类还有用于清除数组元素值的 Clear、复制数组的 Copy、对数据进行排序的 Sort、反转数组元素顺序的 Reverse、从左至右查找数组元素的 IndexOf、从右至左查找数组元素的 LastIndexOf、更新数组长度的 Resize 等方法。

2.4 常量与变量

2.4.1 常量

常量是指在程序中，值保持不变的量。一个数据在程序内频繁地使用，而且保持不变情况下就可以定义成常量。常量包括整型常量、浮点型常量、布尔型常量、字符型常量等，其中 null 常量只有一个值 null，表示对象的引用为空。

常数常量只能在声明时赋值，声明常量的语法格式如下：

[访问修饰符] const 数据类型 变量名=表达式；

例如：

```
const double PI = 3.1415926;
```

2.4.2 变量

在程序运行期间，随时可能产生一些临时数据，这些数据保存在一些内存单元中，每个内存单元都用一个标识符标识。这些内存单元就是变量，定义的标识符是变量名，内存单元中存储的数据是变量的值。一般情况下，声明在类中的称为成员变量，声明在方法中的称为局部变量。

变量必须先声明后使用，声明变量就是给变量指定一个类型和一个名称，声明变量后，编译器会给该变量分配一定大小的内存单元。变量可以在声明时赋值，也可以在运行时赋值。声明变量的语法格式如下：

[访问修饰符] 数据类型 变量名[=表达式]；

例如：

```
int x=1,y;
y=x+15;
```

上面的代码中，第 1 行代码定义了两个变量 x 与 y，相当于分配了两个内存单元，在定义变量的同时为变量 x 赋予了一个初始值 1。第 2 行代码为变量 y 赋值，在执行第 2 行代码时，首先取出变量 x 的值，与 15 相加后，将结果赋值给变量 y。

注意：在 C#语言中，要求所有使用的变量遵循"先定义，后使用"的规则。只有定义了变量，编译系统才会根据变量所属的数据类型分配相应的内存空间，并且编译系统会检查在程序中对该变量进行的运算是否合法。C#的变量名是一种标识符，应该符合标识符的命名规则。

2.4.3 变量的作用域

变量的作用域是指变量的使用范围。在程序中，变量一定会定义在某一对花括号中，该花括号所包含的代码区域就是这个变量的作用域。

例如，在下面代码段中，变量 x 的作用域是第 2～10 行，变量 y 的作用域是第 5～8 行，变量 i 是在循环语句中声明的变量，只存在于该循环体内，即第 5～8 行。

```
1   static void Main(string[] args)
2   {
3       int x = 4;
4       for (int i = 0;i<5;i++)
5       {
6           int y = 3;
7           ...
8       }
9       ...
10  }
```

2.5 运算符

运算符是一种专门用来处理数据运算的特殊符号，数据变量结合运算符形成完整的程序运算语句。运算符用于执行程序代码运算，会针对一个以上操作数项目进行运算。运算符大致可以分为 4 种类型：算术运算符、赋值运算符、比较运算符和逻辑运算符。

2.5.1 算术运算符

算术运算符，就是用来处理四则运算和模（求余）运算的符号，这是最简单也最常用的符号，尤其是数字的处理，几乎都会使用到算术运算符号。C#算术运算符如表 2-6 所示。

表 2-6 C#算术运算符列表

名　　称	符号	说　　明
加法运算符	+	双目运算符，即应有两个量参与加法运算，如 a+b，4+8 等。具有左结合性
减法运算符	−	双目运算符。但"−"也可作负值运算符，此时为单目运算，如-x，-5 等具有左结合性
乘法运算符	*	双目运算符，具有左结合性
除法运算符	/	双目运算符，具有左结合性。参与运算量均为整型时，结果也为整型，舍去小数。如果运算量中有一个是实型，则结果为双精度实型
求余运算符（模运算符）	%	双目运算符，具有左结合性。要求参与运算的量均为整型，不能应用于 float 或 double 类型。求余运算的结果等于两数相除后的余数，整除时结果为 0

【例 2-8】 算数运算符。
【示例代码：chapter02\Solution1\ArithmeticOperators】
程序代码如下：

```
namespace ArithmeticOperators
{
    class Program
    {
        static void Main(string[] args)
```

```
        {
            int a = 100, b = 55;
            Console.WriteLine("a + b=" + (a + b));
            Console.WriteLine("a - b=" + (a - b));
            Console.WriteLine("a * b=" + (a * b));
            Console.WriteLine("a / b=" + (a / b));
            Console.WriteLine("a % b=" + (a % b));
            Console.ReadLine();
        }
    }
}
```

程序运行结果如图 2-8 所示。

图 2-8　ArithmeticOperators 项目程序运行结果

注意：

① 当操作数的类型不同时，如"float result1=6.5+9;"，C#编译系统会自动进行类型转换，先将第二个操作数转换为 float 类型，然后再进行加法运算，整个表达式的结果为 float 类型。

② 在除法运算过程中，默认的返回值类型与精度最高的操作数类型相同。例如，"int result2=7/2;"运算结果为 3，"float result3=7.0/2;"运算结果为 3.5。

2.5.2　赋值运算符

赋值运算符为变量、属性、事件等元素赋新值。赋值运算符主要有=、+=、-=、/=、*=、%=、&=、|=、<<=、>>=和^=运算符。赋值运算符的左操作数必须是变量、属性访问、索引器访问或事件访问类型的表达式，如果赋值运算符两边的操作数的类型不一致，就需要首先进行类型转换，然后再赋值。

在使用赋值运算符时，右操作数表达式所属的类型必须可隐式转换为左操作数所属的类型，运算将右操作数的值赋给左操作数指定的变量、属性或索引器元素。

C# 赋值运算符及其运算规则如表 2-7 所示。

表 2-7　C#赋值运算符及其运算规则

名　　称	运　算　符	运　算　规　则	意　　义
赋值	=	将表达式赋值给变量	将右边的值给左边
加赋值	+=	x+=y	x=x+y

续表

名 称	运 算 符	运 算 规 则	意 义
减赋值	-=	x-=y	x=x-y
除赋值	/=	x/=y	x=x/y
乘赋值	*=	x*=y	x=x*y
模赋值	%=	x%=y	x=x%y
位与赋值	&=	x&=y	x=x&y
位或赋值	|=	x|=y	x=x|y
右移赋值	>>=	x>>=y	x=x>>y
左移赋值	<<=	x<<=y	x=x<<y
异或赋值	^=	x^=y	x=x^y

【例 2-9】 赋值运算符。

【示例代码：chapter02\Solution1\AssignmentOperators】

程序代码如下：

```
namespace AssignmentOperators
{
    class Program
    {
        static void Main(string[] args)
        {
            int a = 2, b = 9, c;
            c = a;
            Console.WriteLine("c=" + c);
            c += a;
            Console.WriteLine("c += a,c=" + c);
            c -= a;
            Console.WriteLine("c -= a,c=" + c);
            c *= a;
            Console.WriteLine("c *= a,c=" + c);
            c /= a;
            Console.WriteLine("c /= a,c=" + c);
            c %= b;
            Console.WriteLine("c %= b,c=" + c);
            c &= b;
            Console.WriteLine("c &= b,c=" + c);
            c |= b;
            Console.WriteLine("c |= b,c=" + c);
            c >>= a;
            Console.WriteLine("c >>= a,c=" + c);
            c <<= a;
            Console.WriteLine("c <<= a,c=" + c);
            c ^= a;
```

```
            Console.WriteLine("c ^= a,c=" + c);
            Console.ReadLine();
        }
    }
}
```

程序运行结果如图 2-9 所示。

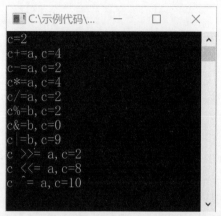

图 2-9　AssignmentOperators 项目程序运行结果

注意：

① 在赋值运算中，赋值号右边可以是变量、常量或表达式，赋值号左边只能是变量。

② 在赋值运算中，如果赋值号两边的数据类型不同，则系统将自动先将赋值号右边的类型转换为左边的类型再赋值。

③ 在赋值运算中，不能把右边数据长度更大的数据类型自动转换，并赋值给左边数据长度更小的数值类型。

例如：

short a = 3,b = 2;
short c = a + b;

赋值语句错误，因为变量 a 和变量 b 虽然都是 short 型，但在进行加法运算时，首先都将被转换为 int 型，int 型的结果不能赋给 short 型的变量。

2.5.3　比较运算符

比较运算符可以完成两个操作数的比较运算之后返回一个代表运算结果的布尔值。C# 比较运算符如表 2-8 所示。

表 2-8　C# 比较运算符

比较运算符	说　　明	比较运算符	说　　明
==	等于	!=	不等于
>	大于	>=	大于或等于
<	小于	<=	小于或等于

【例 2-10】 比较运算符。

【示例代码：chapter02\Solution1\ComparisonOperators】

程序代码如下：

```
namespace ComparisonOperators
{
    class Program
    {
        static void Main(string[] args)
        {
            Decimal a = 1996, b = 1997, c = 1996;
            Console.WriteLine("a == b," + (a == b));
            Console.WriteLine("a > b," + (a > b));
            Console.WriteLine("a < b," + (a < b));
            Console.WriteLine("a != c," + (a != c));
            bool result1 = a >= c;
            bool result2 = a <= c;
            Console.WriteLine("a >= c," + result1);
            Console.WriteLine("a <= c," + result2);
            Console.ReadLine();
        }
    }
}
```

程序运行结果如图 2-10 所示。

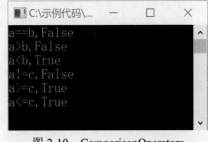

图 2-10 ComparisonOperators 项目程序运行结果

2.5.4 逻辑运算符

逻辑运算符对两个表达式执行布尔逻辑运算。C#中的逻辑运算符大体可以分为按位逻辑运算符和布尔逻辑运算符。

按位逻辑运算符是对两个整数表达式的相应位执行布尔逻辑运算。有效的整型数是有符号或无符号的 int 和 long 类型，它们对每一位执行布尔计算并返回兼容的整数结果。

1. 按位"与"运算符

按位"与"运算符（&）比较两个整数表达式的相应位。当两个整数的对应位都是 1 时，返回相应的结果位是 1；当两个整数的相应位都是 0 或其中一个位是 0 时，则返回相应的结果位是 0。

2. 按位"或"运算符

按位"或"运算符（|）用于比较两个整数的相应位。当两个整数的对应位有一个是 1 或都是 1 时，返回相应的结果位是 1；当两个整数的相应位都是 0 时，则返回相应的结果位是 0。

3. 按位"异或"运算符

按位"异或"运算符(^)用于比较两个整数的相应位。当两个整数的对应位有一个是 1 而另一个是 0 时,返回相应的结果位是 1;当两个整数的相应位都是 1 或都是 0 时,则返回相应的结果位是 0。

4. 布尔"与"运算符

布尔"与"运算符(&)用于计算两个布尔表达式。当两个布尔表达式的结果都是真时,则返回真;否则,返回假。

5. 布尔"或"运算符

布尔"或"运算符(|)用于计算两个表达式的结果。当两个布尔表达式中有一个表达式返回真时,则结果为真;当两个布尔表达式的计算结果都是假时,则结果为假。

6. 布尔"异或"运算符

布尔"异或"运算符(^)用于计算两个布尔表达式的结果,只有当其中一个表达式是真而另外一个表达式是假时,该表达式返回的结果才是真;当两个表达式的计算结果都是真或都是假时,则返回的结果为假。

【例 2-11】 逻辑运算符。

【示例代码:chapter02\Solution1\LogicalOperators】

程序代码如下:

```csharp
namespace LogicalOperators
{
    class Program
    {
        static void Main(string[] args)
        {
            int a = 1, b = 85, c = 20;
            bool d = true;
            Console.WriteLine("a & b=" + (a & b));
            Console.WriteLine("a | b=" + (a | b));
            Console.WriteLine("a ^ b=" + (a ^ b));
            Console.WriteLine("d & (c < 30)," + (d & (c < 30)));
            Console.WriteLine("d | (c < 30)," + (d | (c < 30)));
            Console.WriteLine("d ^ (c < 30)," + (d ^ (c < 30)));
            Console.ReadLine();
        }
    }
}
```

程序运行结果如图 2-11 所示。

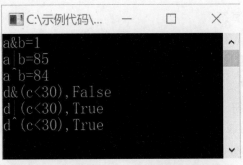

图 2-11　LogicalOperators 项目程序运行结果

2.5.5　运算符优先级

表达式中运算符的计算顺序由运算符的优先级和结合性决定。当表达式包含多个运算符时，运算符的优先级控制各运算符的计算顺序，即高优先级的先计算，低优先级的后计算。表 2-9 给出了 C# 运算符的结合性和优先级。

表 2-9　C# 运算符的结合性和优先级

运算符	结合性	优先级次序
()	从左至右	高
++、--、!	从右至左	
*、/、%	从左至右	
+、-	从左至右	
<、<=、>、>=	从左至右	
==、!=	从左至右	
&&	从左至右	
||	从左至右	
=、+=、*=、/=、%=、-=	从右至左	低

当运算符的优先级相同时，运算符的顺序结合性控制运算符的执行顺序。除了赋值运算符外，所有的二元运算符都是从左向右结合的，即从左向右执行运算。优先级和顺序结合性都可以用括号控制，即括号内的运算优先进行。

2.6　流程控制语句

一个应用程序由许多条语句构成，通常情况下，程序会按先后顺序执行每条语句，如果需要改变程序中语句的执行顺序或重复执行某段程序，就需要使用流程控制语句。常见的流程控制语句包括选择结构语句、循环结构语句和跳转语句。

2.6.1　选择结构语句

选择结构语句也称为条件语句，此类语句根据条件是否成立而控制执行不同的程序段，从而实现程序的分支结构。选择结构语句分为 if 条件语句和 switch 条件语句。

1. if 条件语句

if 条件语句分为 3 种语法格式，分别为 if 语句、if…else 语句和 if…else if…else 语句，每种格式都有其自身特点，适用于解决不同的选择结构问题。

（1）if 语句。

if 语句是指如果满足某种条件，就执行某种处理。if 语句的语法格式如下：

```
if(判断条件)
{
    执行语句;
}
```

上述格式中，判断条件是一个布尔值。当判断条件为 true 时，才会执行花括号中的执行语句。if 语句的执行流程如图 2-12 所示。

【例 2-12】 如果变量 x 的值小于 15，则 x 值增加 1。

【示例代码：chapter02\Solution1\ConditionalStatement\If1】

程序代码如下：

```csharp
namespace ConditionalStatement
{
    class If1
    {
        static void Main(string[] args)
        {
            int x = 9;
            if (x < 15)
            {
                x++;
            }
            Console.WriteLine("x=" + x);
            Console.ReadLine();
        }
    }
}
```

程序运行结果如图 2-13 所示。

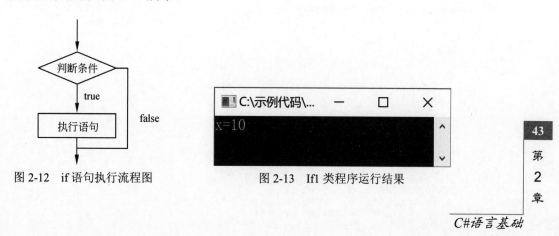

图 2-12 if 语句执行流程图　　　　图 2-13 If1 类程序运行结果

【分析】

本实例为单分支结构，变量 x 值为 9，满足小于 15 的条件，因此执行语句块 x++，即 x 值加 1，输出结果为 10。

（2）if…else 语句。

if…else 语句是指如果满足某种条件，就执行某种处理，否则就执行另一种处理。if…else 语句的语法格式如下：

```
if(判断条件)
{
    执行语句1;
}
else
{
    执行语句2;
}
```

上述格式中，判断条件是一个布尔值。当判断条件为 true 时，执行 if 后面花括号中的执行语句 1；当判断条件为 false 时，执行 else 后面花括号中的执行语句 2。if…else 语句的执行流程如图 2-14 所示。

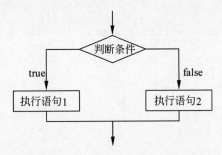

图 2-14　if…else 语句的执行流程图

【例 2-13】　判断奇偶数。

【示例代码:~chapter02\Solution1\ConditionalStatement\If2】

程序代码如下：

```
namespace ConditionalStatement
{
    class If2
    {
        static void Main(string[] args)
        {
            int num = 45;
            if (num % 2 == 0)
            {
```

```
            Console.WriteLine(num + "是一个偶数");
        }
        else
        {
            Console.WriteLine(num + "是一个奇数");
        }
        Console.ReadLine();
    }
  }
}
```

程序运行结果如图 2-15 所示。

【分析】

本实例为双分支结构，判断 num 除 2 的余数是否为 0，如果为 0，则为偶数，否则为奇数。变量 num 值为 45，除 2 余数不为 0，因此结果为奇数。

图 2-15　If2 类程序运行结果

（3）if…else if…else 语句。

if…else if…else 语句用于对多个条件进行判断，进行多种不同的处理。if…else if…else 语句的语法格式如下：

```
if(判断条件1)
{
    执行语句1;
}
else if(判断条件2)
{
    执行语句2;
}
…
else if(判断条件n)
{
    执行语句n;
}
else
{
    执行语句n+1;
}
```

上述格式中，判断条件是一个布尔值。当判断条件 1 为 true 时，执行 if 后面花括号中的执行语句 1；当判断条件 1 为 false 时，会继续执行判断条件 2，如果为 true 则执行语句 2，以此类推。如果所有的判断条件都为 false，则执行 else 后面花括号中的执行语句 n+1。if…else if…else 语句的执行流程如图 2-16 所示。

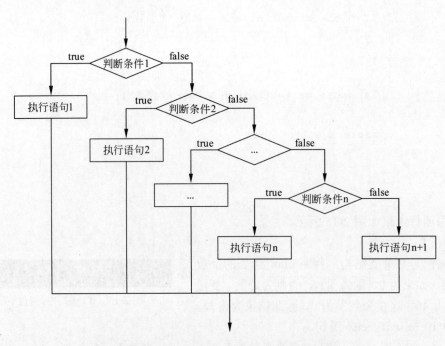

图 2-16 if…else if…else 语句的执行流程图

【例 2-14】 学生成绩等级评定。
【示例代码：chapter02\Solution1\ConditionalStatement\ If3】
程序代码如下：

```
namespace ConditionalStatement
{
    class If3
    {
        static void Main(string[] args)
        {
            int score = 56;
            string grade = "";
            if (score >= 90)
            {
                grade = "优秀";
            }
            else if (score >= 80)
            {
                grade = "良好";
            }
            else if (score >= 70)
            {
                grade = "中等";
            }
```

```
        else if (score >= 60)
        {
            grade = "及格";
        }
        else
        {
            grade = "不及格";
        }
        Console.WriteLine("该生的成绩是："+grade);
        Console.ReadLine();
    }
  }
}
```

程序运行结果如图 2-17 所示。

图 2-17 If3 类程序运行结果

【分析】

本实例为一个多分支结构，首先计算表达式 score >= 90，返回值如果为 true，则执行语句块"grade = "优秀";"即为 grade 赋值为"优秀"，否则依次往下计算各表达式的值，直到某个表达式的值为真，则执行相应的语句块。由于 score 的值为 56，即所有表达式的值都为 false，则执行最后的 else 子句的语句块，即结果为"不及格"。

注意：else if 子句不能作为语句单独出现，必须与 if 配对使用，而最后的 else 子句可以省略，即表示当所有条件都不满足时，什么都不做。

2．switch 条件语句

switch 条件语句与 if 条件语句不同，只能针对某个表达式的值做出判断，从而决定程序执行哪段代码。使用嵌套的 if 语句虽然可以实现多个分支结构，但是当判断条件比较多时，使用 if 语句会降低程序的可读性，而 switch 语句语法简洁，能够处理复杂的条件判断。switch 语句的语法格式如下：

```
switch(条件表达式)
{
    case 目标值1:
    执行语句1;
    break;
    case 目标值2:
    执行语句2;
```

```
        break;
        ⋮
        [default :执行语句n]
}
```

上述格式中,在 switch 语句中使用关键字 switch 描述一个条件表达式,使用 case 关键字描述目标值,当表达式的值和某个目标值匹配时,会执行对应 case 下的语句;如果表达式的值和目标值都不相等,则程序执行 default 后的语句;如果没有 default 标记,则跳出 switch 语句执行后面的语句。switch 语句的执行流程图如图 2-18 所示。

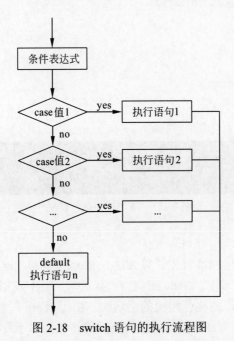

图 2-18　switch 语句的执行流程图

【例 2-15】 学生成绩等级评定。
【示例代码:chapter02\Solution1\ConditionalStatement\ Switch1】
程序代码如下:

```
namespace ConditionalStatement
{
    enum Grade { 优秀, 良好, 中等, 及格, 不及格, none }
    class Switch1
    {
        static void Main(string[] args)
        {
            Grade grade = Grade.none;
            int x = 86;
            switch ((int)(x / 10))
            {
```

```
            case 10:
            case 9:
                grade = Grade.优秀;
                break;
            case 8:
                grade = Grade.良好;
                break;
            case 7:
                grade = Grade.中等;
                break;
            case 6:
                grade = Grade.及格;
                break;
            default:
                grade = Grade.不及格;
                break;
        }
        Console.WriteLine("该生的成绩是: " + grade);
        Console.ReadLine();
    }
}
```

程序运行结果如图 2-19 所示。

图 2-19　Switch1 类程序运行结果

注意：

① 对于多分支结构，使用 if…else 语句的嵌套编写的程序，可读性不高且执行效率偏低，因此建议使用 switch 语句。

② 在 C 和 C++中，允许 switch 语句的 case 标签后不出现 break 语句，但 C#却不同。当语句块不为空时，C#要求每个 case 标签项后必须使用 break 语句或 goto 语句跳出整个 switch 语句，而不允许从一个 case 自动遍历到其他 case。

2.6.2　循环结构语句

循环结构语句是指重复执行某一代码段的语句。循环结构语句的特点是，给定条件成立时，反复执行某代码段，直至条件不成立为止。给定的条件称为循环条件，反复执行的代码称为循环体。C#语言提供了多种循环结构语句，包括 for 语句、while 语句、do…while 语句和 foreach 语句。for 语句、while 语句和 do…while 语句在很多情况下其实是通用的，但一般情况下，如果循环的次数是已知的，推荐选用 for 语句；若循环次数未知，但可以确保至少会执行一次，则推荐使用 do…while 语句；若循环次数完全未知，可以使用 while 语句。

1. for 语句

for 语句的语法格式如下：

```
for(初始表达式；循环条件表达式；循环控制表达式)
{
    循环体语句；
}
```

其中，初始表达式是为循环控制变量赋初始值；循环条件表达式是 bool 型表达式，用于检测循环条件是否成立；循环控制表达式用于更新循环控制变量的值。for 语句的执行流程图如图 2-20 所示。

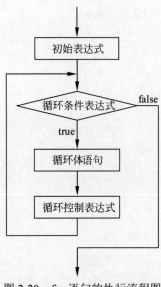

图 2-20　for 语句的执行流程图

【例 2-16】 求 1 到 100 的累加之和。

【示例代码：chapter02\Solution1\LoopStatement\For1】

程序代码如下：

```
namespace LoopStatement
{
    class For1
    {
        static void Main(string[] args)
        {
            int sum = 0;
            for (int i = 1; i <= 100; i++)
            {
                sum += i;
            }
            Console.WriteLine("1到100的累加之和是："+sum);
```

```
            Console.ReadLine();
        }
    }
}
```

程序运行结果如图 2-21 所示。

图 2-21　For1 类程序运行结果

【分析】

本实例是一个典型的需要反复累加计算的数学问题，需要使用循环结构来解决。其中设置一个循环控制变量 i，初始值为 1，再定义一个 sum 变量用于保存累加和。每循环一次，先检查循环控制变量 i 是否小于 100，如果是，则执行 sum += i，把变量 i 的值累加到 sum 变量中，再将 i 的值增加 1；如果不是，则跳出循环。

注意：

① for 循环运行顺序：初始表达式→循环条件表达式→循环体→循环控制表达式。

② 分号不能省略。

③ 语句体只有一条语句时可以省略{}，两句或以上的不能省略。

2．while 语句

while 语句的语法格式如下：

```
while(循环条件)
{
    循环体语句;
}
```

其中，循环条件是 bool 型表达式。当循环条件表达式的值为 true 时，反复执行循环体语句；当表达式的值为 false 时，执行 while 语句块后面的语句。while 语句的执行流程图如图 2-22 所示。

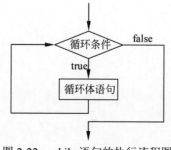

图 2-22　while 语句的执行流程图

【例2-17】 求1到100的累加之和。

【示例代码：chapter02\Solution1\LoopStatement\While1】

程序代码如下：

```
namespace LoopStatement
{
    class While1
    {
        static void Main(string[] args)
        {
            int i = 1, sum = 0;
            while (i <= 100)
            {
                sum += i;
                i++;
            }
            Console.WriteLine("1到100的累加之和是：" + sum);
            Console.ReadLine();
        }
    }
}
```

程序运行结果如图2-23所示。

图2-23　While1类程序运行结果

注意：

① while循环运行顺序：循环条件表达式→循环体。

② while循环体中的语句必须修改循环条件的值，否则会形成死循环。

3．do…while语句

do…while语句的语法格式如下：

```
do
{
    循环体语句;
}
while(循环条件);
```

其中，do…while语句首先执行一次循环体语句序列，再进行条件判断。如果为true，再次执行循环体语句；否则，退出循环，执行后面的语句。do…while 语句的执行流程图如图 2-24 所示。

【例 2-18】 求 1 到 100 的累加之和。

【示例代码：chapter02\Solution1\LoopStatement\DoWhile1】

程序代码如下：

```
namespace LoopStatement
{
    class DoWhile1
    {
        static void Main(string[] args)
        {
            int i = 1, sum = 0;
            do
            {
                sum += i;
                i++;
            }
            while (i <= 100);
            Console.WriteLine("1到100的累加之和是："+sum);
            Console.ReadLine();
        }
    }
}
```

程序运行结果如图 2-25 所示。

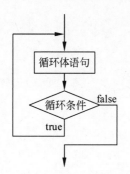

图 2-24　do…while 语句的执行流程图

图 2-25　DoWhile1 类程序运行结果

注意：

① do…while 循环运行顺序：循环体→循环条件表达式。

② do…while 与 while 的区别：前者至少执行一次，先执行后判断。后者先判断后执行，条件为假不会执行；条件为真时结果一样，条件为假时 do 语句执行一次。

③ while 后的分号不能省略。

4. foreach 语句

foreach 语句用于循环访问集合中的元素,但不能更改集合中的元素,循环变量是一个只读变量,且只能用于 foreach 循环体内。foreach 语句的语法格式如下:

foreach(类型 循环变量 in 集合)
{
　　循环体语句;
}

其中,类型和循环变量用于声明迭代变量,迭代变量相当于一个范围覆盖整个循环体语句的局部变量,在 foreach 语句执行期间,迭代变量表示当前正在为其执行迭代的集合元素。foreach 语句的执行流程图如图 2-26 所示。

foreach 语句的执行过程如下:
(1)自动指向数组或集合中的第一个元素。
(2)判断该元素是否存在,如果不存在,则结束循环。
(3)把该元素的值赋值给循环变量。
(4)执行循环体语句块。
(5)自动指向下一个元素,之后从步骤(2)开始重复执行。

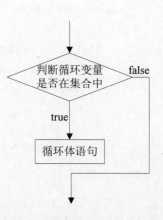

图 2-26　foreach 语句的执行流程图

【例 2-19】统计字符串中字符 s 出现的次数。

【示例代码:chapter02\Solution1\LoopStatement\Foreach1】

程序代码如下:

```
namespace LoopStatement
{
    class Foreach1
    {
        static void Main(string[] args)
        {
            string s = "This is Visual studio 2022!";
            int num = 0;
            foreach (char i in s)
            {
                if (i == 's')
                    num++;
            }
            Console.WriteLine("字符串{0}中一共包含{1}个s。",s,num);
            Console.ReadLine();
        }
    }
}
```

程序运行结果如图 2-27 所示。

图 2-27　Foreach1 类程序运行结果

注意：

① foreach 语句总是遍历整个集合。如果只需要遍历集合的特定部分或需要绕过特定元素时，最好使用 for 语句。

② foreach 语句总是从集合中的第一个元素遍历到最后一个元素，如果需要反向遍历，最好使用 for 语句。

③ 如果循环体需要知道元素索引，而不仅是元素值时，则必须使用 for 语句。

④ foreach 语句读出的元素变量是一个只读变量，不能进行修改，如果需要修改数组元素，必须使用 for 语句。

2.6.3　跳转语句

跳转语句用于实现程序执行过程中有目的的跳转，在 C#中的跳转语句包括 break 语句、continue 语句、return 语句和 goto 语句。其中以 break 语句和 return 语句最为常见，而 goto 语句一般不推荐使用，因为它可能导致程序难以阅读和维护。

1. break 语句

break 语句可直接跳出当前的整个循环，可用在循环和 switch 中。

【例 2-20】　使用 break 语句跳出当前循环。

【示例代码：chapter02\Solution1\SkipStatement\Break1】

程序代码如下：

```
namespace SkipStatement
{
    class Break1
    {
        static void Main(string[] args)
        {
            int a = 1;
            while (a <= 4)
            {
                Console.WriteLine("a=" + a);
                if (a == 3)
                {
                    break;
```

```
            }
            a++;
        }
        Console.ReadLine();
    }
}
```

程序运行结果如图 2-28 所示。　　　　　图 2-28　Break1 类程序运行结果

2．continue 语句

continue 语句只能用在循环中，continue 语句不是终止并跳出当前循环，而是终止执行 continue 语句后面的语句，直接回到当前循环的起始处，开始下一次循环。

【例 2-21】　使用 continue 语句实现输出 1～30 中不能被 2 整除的数。

【示例代码：chapter02\Solution1\SkipStatement \ Continue1】

程序代码如下：

```
namespace SkipStatement
{
    class Continue1
    {
        static void Main(string[] args)
        {
            for (int i = 1; i <= 30; i++)
            {
                if (i % 2 == 0)
                {
                    continue;
                }
                Console.Write(i + " ");
            }
            Console.ReadLine();
        }
    }
}
```

程序运行结果如图 2-29 所示。

图 2-29　Continue1 类程序运行结果

注意：continue 语句容易与 break 语句混淆，也是一个用于循环控制的语句，其作用不是退出整个循环，只是将程序的执行流程提前跳转到下一次循环，执行流程仍然在循环内，而 break 语句使得程序的执行流程从循环内跳转到循环外。例如，本实例中，凡是满足被 2 整除的数值都没有被输出，其他数值都被正常输出，表明程序遇到 continue，并未跳出循环外，只是略过了某些满足条件的循环而已。

3．return 语句

return 语句用于终止方法的执行，并将返回值返回给调用方法。

【例 2-22】 使用 return 语句实现求两个数和的方法。

【示例代码：chapter02\Solution1\SkipStatement \Return1】

程序代码如下：

```
namespace SkipStatement
{
    class Return1
    {
        static void Main(string[] args)
        {
            Console.WriteLine("34 + 57=" + Add(34, 57));
            Console.ReadLine();
        }
        static int Add(int x, int y)
        {
            return x + y;
        }
    }
}
```

程序运行结果如图 2-30 所示。

图 2-30　Return1 类程序运行结果

4．goto 语句

goto 语句可将程序直接跳转到标记语句。goto 语句只能在一个方法体中进行语句跳转，且同一个方法体中标记名是唯一的。

【例 2-23】 使用 goto 语句实现跳转到标记语句，求解 32×15 的正确答案。

【示例代码：chapter02\Solution1\SkipStatement \Goto1】

程序代码如下：

```
namespace SkipStatement
{
    class Goto1
    {
        static void Main(string[] args)
        {
            Console.WriteLine("32*15的值为？请选择正确答案！");
            Console.WriteLine("a. 470");
            Console.WriteLine("b. 480");
            Console.WriteLine("c. 490");
            Console.WriteLine("d. 500");
            string re = Console.ReadLine();
            switch (re)
            {
                case "a":
                    goto error;
                case "b":
                    goto right;
                case "c":
                    goto error;
                case "d":
                    goto error;
                default:
                    goto def;
            }
            //标记答案正确
            right:
                {
                    Console.WriteLine("答案正确！");
                    Console.ReadLine();
                }
            //标记答案错误
            error:
                {
                    Console.WriteLine("答案错误！");
                    Console.ReadLine();
                }
            //标记选项不存在
            def:
                {
                    Console.WriteLine("选项不存在！");
```

```
                Console.ReadLine();
            }
        }
    }
}
```

程序运行结果如图 2-31 所示。

图 2-31　Goto1 类程序运行结果

2.7　数据类型转换

在程序中，当把一种数据类型的值赋给另一种数据类型的变量时，需要进行数据类型转换。根据转换方式的不同，数据类型转换可以分为自动转换和强制转换，同时，某些值类型之间也可以采用 Convert 类提供的静态方法进行转换。

1. 自动转换

自动转换又称为隐式转换，指的是两种数据类型在转换的过程中不需要显式地进行声明。自动转换一般在不同类型的数据进行混合运算时发生，当编译器能判断转换的类型，而且转换不会带来精度的损失时，C#语言编译器会自动进行转换。自动转换一般是安全的，不会造成数据溢出或丢失等问题。

要实现自动类型转换，必须同时满足两个条件：

（1）两种数据类型彼此兼容。

（2）目标类型的取值范围大于源类型的取值范围。

进行自动类型转换时，遵循以下规则：

（1）如果参与运算的数据类型不相同，则先转换成同一类型，然后进行运算。

（2）转换时按数据长度增加的方向进行，以保证精度不降低。例如，int 型和 long 型运算时，先把 int 型数据转换成 long 型后再进行运算。

（3）所有的浮点运算都是以双精度进行的，即使仅有 float 单精度运算的表达式，也要先转换成 double 型，再进行运算。

（4）byte 型和 short 型参与运算时，必须先转换成 int 型。

（5）char 型可以隐式转换成 ushort、int、uint、long、ulong、float、double 或 decimal 型，但不存在从其他类型到 char 类型的隐式转换。

例如：

```
byte x = 9;
int y = x;
```

在上面程序中，不需要特殊声明，byte 类型的变量 x 就直接转换成 int 类型，因为 x 是一个 byte 字节型，占 8 位，y 是一个 int 整型，占 64 位，编译器自动转换后，不会损失精度。

以下代码在编译时将出现错误。

```
int x = 9;
unit y = x;
```

虽然 int 类型和 unit 类型都占 32 位，但 uint 类型不能存储负数，因此不能进行自动数据类型转换。

2. 强制转换

强制转换又称为显式转换，指的是两种数据类型之间的转换需要显式地进行声明。当两种类型彼此不兼容，或目标类型取值范围小于源类型时，就需要进行强制转换。强制转换的语法格式如下：

源类型　变量 =（目标类型）值

【例 2-24】 使用强制转换方法，分别将 double 类型强制转换成 int 类型，int 类型强制转换为 Weekdays 类型。

【示例代码：chapter02\Solution1\TypeConversion\Convert1】

程序代码如下：

```
namespace TypeConversion
{
    class Convert1
    {
        static void Main(string[] args)
        {
            double x = 123.45;
            int y = (int)x;
            Weekdays w = (Weekdays)2;
            Console.WriteLine(y);
            Console.WriteLine(y.GetType());
            Console.WriteLine(w);
            Console.WriteLine(w.GetType());
            Console.Read();
        }
    }
    enum Weekdays { 星期日, 星期一, 星期二, 星期三, 星期四, 星期五, 星期六 };
}
```

程序运行结果如图 2-32 所示。

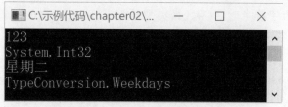

图 2-32　Convert1 类程序运行结果

3. Convert 类的方法

Convert 类可以将一个基本数据类型转换成另一个基本数据类型。

【例 2-25】　使用 Convert 类实现将 string 类型转换为 int 类型。

【示例代码：chapter02\Solution1\TypeConversion\Convert2】

程序代码如下：

```
namespace TypeConversion
{
    class Convert2
    {
        static void Main(string[] args)
        {
            Console.WriteLine("请输入一个加数");
            string a = Console.ReadLine();
            Console.WriteLine("请输入一个被加数");
            string b = Console.ReadLine();
            int x = Convert.ToInt32(a);
            int y = Convert.ToInt32(b);
            int re = x + y;
            Console.WriteLine("两数和是:" + re);
            Console.ReadLine();
        }
    }
}
```

图 2-33　Convert2 类程序运行结果

程序运行结果如图 2-33 所示。

注意：

① 待转换数据不是单个变量时，类型和待转换数据都必须加圆括号。例如，(int)(x+y)表示把 x+y 的结果转换为 int 型。

② 无论是强制转换还是自动转换，都只是临时性转换，不会改变变量声明时对该变量定义的类型。

③ 当被转换的目标为字符串时，C#内置的简单类型均自带 Parse()方法，调用该方法可自动解析字符串，并转换为指定的数据类型。例如：

```
int x = int.Parse("2016.9");
```

④ 将变量转换为字符串时，C#数据类型均带有 ToString()方法，调用该方法可将数据类型转换为对应的字符串。例如：

```
int x = 2016;
string y = a.ToString();
```

2.8 异常处理

程序在运行时可能会发生非正常情况，如磁盘空间不足、用户输入错误、被操作的文件不存在等。针对这种情况，C#程序引入了异常处理机制。采用异常处理可以解决一部分错误，这对于提高程序的可靠性有非常重要的作用。

在C#中提供了大量的异常类，这些类都继承自 Exception 类，每个异常类都代表一个指定的异常类型，常用的异常类如表 2-10 所示。

表 2-10 常用的异常类

异 常 类	描 述
Exception	所有异常对象的基类
SystemException	运行时产生所有错误的基类
IndexOutOfRangeException	当一个数组的下标超出范围时引发
NullReferenceException	当一个空对象被引用时引发
InvalidOperationException	当对方法的调用且对对象的当前状态无效时引发
ArgumentException	所有参数异常的基类
ArgumentNullException	当参数为空的情况下引发
ArgumentOutOfRangeException	当参数不在一个给定范围之内时引发
InteropException	目标在或发生在 CLR 外面环境中异常的基类
ComException	包含 COM 类 HRESULT 信息的异常
SEHException	封装 Win32 结构异常处理信息的异常
SqlException	封装了 SQL 操作异常

为了更好地展示异常信息，每个异常对象中都包含一些只读属性，这些属性可以描述异常的信息，通过这些属性可以更准确地找到异常出现的原因，具体如表 2-11 所示。

表 2-11 异常对象的常用只读属性

名 称	类 型	描 述
Message	string	该属性含有解释异常原因的消息
Source	string	该属性含有异常起源所在程序集的名称
StackTrace	string	该属性含有描述异常发生的位置信息
HelpLink	string	该属性为异常原因信息提供 URL 或 URN
InnerException	Exception	如果当前异常由另一个异常引起，则该属性包含前一个异常引用

1. try…catch 语句

try…catch 语句提供了捕获和处理指定异常的方法，它的语法格式如下：

```
try
{
    语句
```

```
catch [(异常类型 异常对象)]
{
    异常处理
}
```

当程序正常运行时,执行的是 try 块的语句,如果 try 后的任何语句发生异常,程序都会转移到 catch 块的语句处理异常。

【例 2-26】 使用 try…catch 语句捕获访问整型数组 array 时产生索引越界异常,并输出提示信息。

【示例代码:chapter02\Solution1\ExceptionStatement\TryCatch】

程序代码如下:

```
namespace ExceptionStatement
{
    class TryCatch
    {
        static void Main(string[] args)
        {
            int[] array = { 1, 2, 3, 4, 5 };
            try
            {
                for (int i = 0; i <= array.Length; i++)
                {
                    Console.Write(array[i].ToString()+" ");
                }
            }
            catch (Exception ex)
            {
                Console.Write(ex.Message.ToString());
            }
            Console.ReadLine();
        }
    }
}
```

程序运行结果如图 2-34 所示。

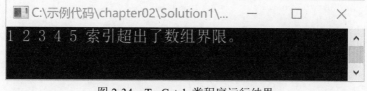

图 2-34 TryCatch 类程序运行结果

注意：

① catch 后面也可以不带任何内容，既没有指定异常类型，也没有指定异常对象名，此 catch 语句称为通用 catch 语句。一个 try 块只能存在一条通用 catch 语句，而且它必须是 try 块中的最后一条 catch 语句。

② 当 catch 语句中指定一个异常类型时，此类型必须为 System.Exception 或其派生类。

③ 当 catch 语句中同时指定了异常类型和异常对象时，该对象代表当前正在被处理的异常，可以在 catch 语句块内部使用该对象，但不能赋值。

④ 可以有一个或多个 catch 块，但由于 catch 语句是依照它们出现的次序进行检查，进而确定哪一个 catch 处理异常，所以，如果某一 catch 语句指定的类型同它之前的 catch 语句指定的类型一致，或由此类型派生而来，那么运行时会出现错误。

⑤ try 和 catch 后的一对花括号是必要的，即使代码块中只有一条语句。

2. try…catch…finally 语句

try…catch…finally 语句提供了捕获异常、处理异常并继续执行应用程序的方法，它的语法格式如下：

```
try
{
    语句
}
catch [(异常类型 异常对象)]
{
    异常处理
}
finally
{
    语句
}
```

try…catch…finally 语句与 try…catch 语句的区别在于：不论程序执行时是否出现异常，也不论是否有 catch 语句处理异常，finally 语句块都会执行。

【例 2-27】 使用 try…catch…finally 语句捕获访问整型数组 array 时产生索引越界异常。

【示例代码：chapter02\Solution1\ExceptionStatement\TryCatchFinally】

程序代码如下：

```
namespace ExceptionStatement
{
    class TryCatchFinally
    {
        static void Main(string[] args)
        {
            int[] array = { 1, 2, 3, 4, 5 };
            try
```

```
            {
                for (int i = 0; i <= array.Length; i++)
                {
                    Console.Write(array[i].ToString() + " ");
                }
            }
            catch (Exception ex)
            {
                Console.WriteLine(ex.Message.ToString());
            }
            finally
            {
                Console.WriteLine("不论是否有异常都会运行finally语句块！");
            }
            Console.ReadLine();
        }
    }
}
```

程序运行结果如图 2-35 所示。

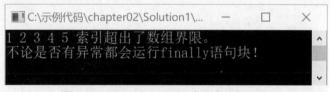

图 2-35　TryCatchFinally 类程序运行结果

注意：
① finally 语句块中不允许出现 return 语句。
② 可以省略 catch 块，即 try…finally 结构，该结构不对异常进行处理。

3．throw 语句

当程序中出现异常时，还可以使用 throw 语句抛出异常对象，该异常既可以是预定的异常类，也可以是自定义的异常类。它的语法格式如下：

```
throw new 异常 (异常描述);
```

【例 2-28】 使用 throw 语句抛出异常。
【示例代码：chapter02\Solution1\ExceptionStatement\Throw】
程序代码如下：

```
namespace ExceptionStatement
{
    class Throw
    {
```

```csharp
static void Main(string[] args)
{
    string str = "hello";
    try
    {
        int i = ConvertStringToInt(str);
        Console.WriteLine(i.ToString());
    }
    catch(Exception ex)
    {
        Console.WriteLine(ex.Message.ToString());
    }
    Console.ReadLine();
}
static int ConvertStringToInt(string str)
{
    try
    {
        return Convert.ToInt32(str);;
    }
    catch
    {
        throw new FormatException("转换错误异常");
    }
}
```

程序运行结果如图 2-36 所示。

图 2-36 Throw 类程序运行结果

习题2

第 3 章　面向对象程序设计

学习目标：
- 理解面向对象的概念，理解类与对象的区别；
- 掌握类的声明和实例化方法；
- 掌握属性、方法以及构造方法的定义；
- 了解方法的重载以及方法的高级参数；
- 了解访问修饰符；
- 掌握静态类和静态成员的访问方法；
- 理解和掌握面向对象的基本特征。

3.1　面向对象简介

面向对象不仅是一项具体的软件开发技术，也是一种符合人类思维习惯的编程思想。现实生活中存在各种形态不同的事物，这些事物之间存在着各种各样的联系。面向对象编程（object-oriented programming，OOP）就是利用对象建模技术分析目标问题，抽象出相关对象的共性，并对共性进行分类及分析各类之间的关系，同时使用类描述同一类问题。

面向对象中类的定义充分体现了抽象数据类型的思想，基于类的体系结构可以把程序的修改局部化，特别是一旦系统功能需要修改时，只要修改类中间的某些操作，而类所代表的对象基本不变，保持整个系统仍然稳定。

3.2　类　与　对　象

面向对象的编程思想力图使程序对事物的描述与该事物在现实中的形态保持一致，为了做到这一点，在面向对象的思想中提出了两个概念，即类和对象。

类是对某类事物的抽象描述，对象是该类事物的某一个实体，对象会被分配物理内存。以人类为例，每个人都可以看作是一个对象，如图 3-1 所示。类是描述多个对象的共同特征，是具有相同特性（属性）和行为（方法）的一组对象的集合。例如，人类需要描述的特征和行为包括名字、年龄、身高以及说话、唱歌等。对象用于描述现实中的实体，对象是类的实例化。例如，"李雷"是一个具体的人，是一个对象，应该具有人类的属性和方法等。

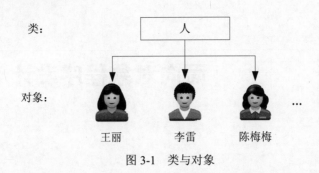

图 3-1 类与对象

3.2.1 类的声明

为了在程序中创建对象,首先需要声明一个类,用于描述一组对象的特征和行为。类中可以定义字段、属性、方法等成员。定义在类中的变量称为字段,字段用于在类中存储数据,属性用于描述对象的特征,而方法用于描述对象的行为。

注意:如果类的声明中没有指定字段的初始值,使用对象时也没有给字段赋值,则编译时会自动赋予其类型的默认值并发出警告。声明类的语法格式如下:

```
[访问修饰符] class 类名称 [:基类或接口]
{
    类成员定义
}
```

【**例 3-1**】 声明一个 Person 类。

【**示例代码**:chapter03\Solution1\ClassAndObject\Person】

```
class Person
    {
        private string _name;
        public string Name
        {
            get{return _name;}
            set{_name = value;}
        }
        public void Speak()
        {
            Console.WriteLine("大家好,我是" + _name);
        }
    }
```

【**分析**】

本实例中,Person 是类名;_name 是字段;Name 是属性;Name 属性封装了_name 字段;Speak()是方法。

3.2.2 对象的创建与使用

1．对象的创建

在 C#中可以使用 new 关键字创建对象，创建对象的语法格式如下：

类名 对象名称 = new 类名();

例如，创建 Person 类的实例，程序代码如下：

Person p = new Person();

其中，Person p 是声明了一个 Person 类型的变量 p；new Person()用于创建 Person 类的一个实例对象；中间的等号用于将 Person 对象在内存中的地址赋值给变量 p，这样变量 p 就有了 Person 对象的引用。

2．对象的使用

在创建 Person 对象后，可以通过对象的引用访问对象所有的成员，语法格式如下：

对象引用.对象成员

例如：

p.Speak();

3.3 类的数据成员

类的成员包括类的常量、字段、属性、索引器、方法、事件、构造方法等，其中，常量、字段和属性都是与类的数据有关的成员。

1．常量

在第 2 章中已经介绍了常量的概念，在类中的常量成员是一种符号常量，符号常量是由用户根据需要自行创建的常量。在程序设计过程中，这些符号常量可能需要反复使用，如定义圆周率 PI 值为 3.1415926。

2．字段

字段表示类的成员变量，字段的声明方式即变量的声明方式，字段的值代表某个对象数据状态。字段使类具备封装数据的能力，一般情况下，应该将字段声明为 private，然后通过属性或方法访问其内容。

3．属性

在 C#中，为保证类中内部数据的安全，可以使用属性封装字段，首先需要将字段访问级别设为 private，再通过属性的 get 和 set 访问器对字段进行读写操作，具体语法格式如下：

```
[访问修饰符] 数据类型 属性名
{
    get{获得属性的代码;}
    set{设置属性的代码;}
}
```

如果设置读写属性，需要同时使用 get 和 set 访问器；如果设置只读属性，只需要使用 get 访问器，一般用于在构造方法中给属性赋值，在程序运行过程中不能修改该属性的值；如果设置只写属性，需要使用 set 访问器，在程序运行过程中只能写入值而不能读取值；如果设置自动属性，则不需要书写任何属性的代码，也就是在 get 和 set 访问器后面不加花括号，直接加";"即可。

【例 3-2】 定义 ComputerStudent 类的自动属性、只读属性和读写属性。

【示例代码：chapter03\ Solution1\Property】

程序代码如下：

```
namespace Property
{
    class Program
    {
        static void Main(string[] args)
        {
            ComputerStudent s = new ComputerStudent();
            s.Age = 19;
            s.Speak();
            Console.ReadLine();
        }
    }
    public class ComputerStudent
    {
        //自动属性
        public string Name {get; set;}
        //只读属性
        private string _specialtyName = "计算机科学与技术";
        public string SpecialtyName
        {
            get { return _specialtyName; }
        }
        //读写属性
        private int _age;
        public int Age
        {
            get
            {
                return _age;
            }
            set     //检查参数合法性
            {
                if (value <= 0)
                {
                    Console.WriteLine("年龄输入不合法！");
                }
```

```
            else
            {
                _age = value;
            }
        }
    }
    public void Speak()
    {
        Console.WriteLine("我今年"+_age+"岁,专业是: "+_specialtyName);
    }
}
```

程序运行结果如图 3-2 所示。

图 3-2　Property 项目程序运行结果

【分析】

在本实例中,通过字段存储数据,而通过属性完成对字段的访问,通过方法对数据进行操作,将操作得到的结果交给调用方,数据与操作融为一体,这就是封装。

注意:从 C# 3.0 开始出现了一种新的、简洁的属性定义方式,即自动属性。在自动属性中,无须定义一个相应的私有字段,也不必写任何 return 和 value 语句。自动属性虽然简洁,但不能通过属性完成任何更多的复杂逻辑。

3.4　方　　法

在类中自定义的"函数"称为"方法"。方法是表示实现类功能而执行的计算或操作。

3.4.1　方法的定义与调用

每个方法都有一个名称和一个主体。方法名应该是一个有意义的标识符,描述方法的用途;方法主体包含了调用方法时实际执行的语句。定义方法的语法格式如下:

```
[访问修饰符] 返回值类型 方法名(参数列表)
{
    方法体
    [return 返回值;]
}
```

上述语法格式中,需要注意以下几点:

（1）方法的返回类型是指调用方法后返回值的类型，如果没有返回值，则为 void；如果有返回值，则在语句序列中必须使用 return 语句返回一个值，这个值的类型必须与返回类型一致。

（2）参数可有可无，如果有参数，每个参数都应指定数据类型和参数名，参数之间用逗号隔开。方法定义时参数列表中定义的参数称为形式参数（形参），而实际调用时传递的参数称为实际参数（实参）。形参和实参一一对应，个数和数据类型必须一致。

（3）方法调用有 3 种方式：

① 在同一个类中，方法可以直接调用。

② 在其他类中的方法，需要通过类的实例进行调用。

③ 静态方法需要通过类名进行调用。

【例 3-3】 通过定义方法 Add()和 Subtract()实现加减法运算。

【示例代码：chapter03\ Solution1\Method\AddAndSubtract】

程序代码如下：

```
namespace Method
{
    class AddAndSubtract
    {
        static void Main(string[] args)
        {
            int num1 = 8;
            int num2 = 39;
            Console.WriteLine("{0}+{1}={2}", num1, num2, Add(num1, num2));
            Console.WriteLine("{0}-{1}={2}", num1, num2, Subtract(num1,num2));
            Console.ReadLine();
        }
        //定义方法(加法)
        static int Add(int x, int y)
        {
            return x + y;
        }
        //定义方法(减法)
        static int Subtract(int x, int y)
        {
            return x - y;
        }
    }
}
```

程序运行结果如图 3-3 所示。

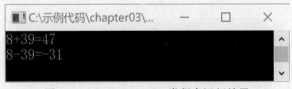

图 3-3　AddAndSubtract 类程序运行结果

【分析】

本实例定义了 Add()方法和 Subtract()方法实现两个整型参数的加法和减法运算,返回值均为整型。在同一个类的方法中,直接调用 Add()方法和 Subtract()方法,传入实参 num1 和 num2 的值,求出返回值并输出。

3.4.2 方法的重载

方法重载是一种操作性多态。当需要在多个不同的实现中对不同的数据执行相同的逻辑操作时,就可以使用重载。例如,Console 类的 WriteLine()方法具有 19 个重载。

方法重载指在同一个类中创建多个同名的方法,但这些方法的参数互不相同,可以是参数类型不同,也可以是参数个数不同。决定方法是否构成重载的 3 个条件如下:

(1)在同一个类中。
(2)方法名相同。
(3)参数列表不同。

【例 3-4】 通过方法重载分别实现对两个整数和 3 个整数的加法程序。

【示例代码:chapter03\Solution1\Method\Overload】

程序代码如下:

```
namespace Method
{
    class Overload
    {
        static void Main(string[] args)
        {
            int num1 = 8;
            int num2 = 39;
            int num3 = 11;
            Console.WriteLine("{0}+{1}={2}", num1, num2, Add(num1, num2));
            Console.WriteLine("{0}+{1}+{2}={3}", num1, num2, num3,Add(num1,
             num2,num3));
            Console.ReadLine();
        }
        //实现两个整数相加
        static int Add(int x, int y)
        {
            return x + y;
        }
        //实现3个整数相加
        static int Add(int x,int y,int z)
        {
            return x+y+z;
        }
    }
}
```

程序运行结果如图 3-4 所示。

图 3-4　Overload 类程序运行结果

【分析】

在本实例中，在 Overload 类中定义了两个求和方法，方法名都是 Add，但一个 Add() 方法有两个参数，另一个 Add() 方法有 3 个参数，即参数列表个数不同，因此实现了方法的重载。在调用 Add() 方法时，通过输入不同的参数，执行不同的方法体。

3.4.3　方法的高级参数

为了让方法的调用更加灵活，C#中引入了高级参数的概念。通过高级参数可以实现两个功能，一是调用方法时允许传入任意个数的参数，而不受形参个数的约束；二是允许在方法中修改方法以外变量的值。高级参数有以下 3 种：

（1）params 参数：用于实现方法接收任意个数的参数。

（2）ref 参数：用于传递参数的引用，而不是参数的值。

（3）out 参数：用于将值从方法体内传到方法体外。

1．params 参数

params 参数在定义时不需要确定参数的个数，可以使用 params 关键字加上数组的方式作为方法的参数，在调用方法时传入任意个数类型相同的参数即可。

【例 3-5】 定义一个使用 params 参数的加法方法。

【示例代码：chapter03\Solution1\Method\Params】

程序代码如下：

```
namespace Method
{
    class Params
    {
        static void Main(string[] args)
        {
            int sum = Add(8, 39, 11);
            Console.WriteLine("8, 39, 11三数之和是：" + sum);
            Console.ReadLine();
        }
        static int Add(params int[] nums)
        {
            int sum = 0;
            for (int i = 0; i < nums.Length; i++)
```

```
            {
                sum += nums[i];
            }
            return sum;
        }
    }
}
```

程序运行结果如图 3-5 所示。

图 3-5 Params 类程序运行结果

【分析】

在静态方法 Add()中使用了 params 参数，参数以整型数组的方式存在，在主方法中调用 Add()方法时，可传入任意个实参。显然，例 3-5 中通过使用 params 参数简化了例 3-4 中多个数求和运算的代码。

注意：

① 一个方法中最多只能出现一个 params。

② params 关键字只能放到所有参数的最后面，即 params 修饰的参数后面不能再有其他参数。

③ 当参数为 params 修饰时，要防止外界传入非法参数，如 null。

2．ref 参数

在使用引用传递时，需要在参数前面加上 ref 关键字，并在调用方法时也使用 ref 关键字，就可以实现当方法执行后，外面的变量值保持方法内修改后的结果。

【例 3-6】 定义一个使用 ref 参数的方法，实现方法内部交换两个参数值。

【示例代码：chapter03\Solution1\Method\Ref】

程序代码如下：

```
namespace Method
{
    class Ref
    {
        static void Main(string[] args)
        {
            int num1 = 8, num2 = 39;
            Exchange(ref num1, ref num2);
            Console.WriteLine("num1 = {0},num2 = {1}",num1 ,num2);
            Console.ReadLine();
        }
```

```
        static void Exchange(ref int num1,ref int num2)
        {
            int temp = num1;
            num1 = num2;
            num2 = temp;
        }
    }
}
```

程序运行结果如图 3-6 所示。

图 3-6 Ref 类程序运行结果

【分析】

在本实例中的 Exchange()方法将两个 ref 参数 num1 与 num2 的值进行调换，在主方法中，调用 Exchange()方法后，变量 num1 与 num2 的值变为方法内调换后的结果。

3．out 参数

out 参数也用于实现引用传递，使用 out 参数可以将值从方法体内传到方法体外。

【例 3-7】 定义一个使用 out 参数的方法，实现参数返回数组中的最大值和最小值。

【示例代码：chapter03\Solution1\Method\Out】

程序代码如下：

```
namespace Method
{
    class Out
    {
        static void Main(string[] args)
        {
            int[] nums = {7,9,32,48,2,97};
            int max, min;
            GetValues(nums, out max, out min);
            Console.WriteLine("最大值是{0}，最小值是{1}",max,min);
            Console.ReadLine();
        }
        static void GetValues(int[] nums, out int max, out int min)
        {
            max = nums[0];
            min = nums[0];
            for (int i = 0; i < nums.Length; i++)
```

```
            {
                if (max < nums[i]) max = nums[i];
                if (min > nums[i]) min = nums[i];
            }
        }
    }
}
```

程序运行结果如图 3-7 所示。

图 3-7　Out 类程序运行结果

【分析】

在本实例中，使用 GetValues()方法求数组 nums 中的最大值 max 与最小值 min，并使用 out 参数将 max 与 min 的值传递给方法体。在主方法中通过调用 GetValues()方法获得实际数组中的最大值 97 和最小值 2。

注意：

① 一般情况下，如果参数前面没有 ref 和 out，则发生调用时，实参的值会被复制给形参，虽然此时形参和实参的值相同，但是它们却占据着不同的内存空间，已不再有任何关联，即在方法里对形参的任何改变都不会影响实参，方法里对形参值的改变只会反映到形参所占的内存空间，并不会影响实参所占据的内存空间。若参数前面有 ref 或 out 修饰时，此时调用，并不是完成实参向形参的复制，而是形参与实参都共享一块内存空间。故在方法内对形参的改变，其实就是对实参的改变，而且这种改变在方法调用结束后仍旧维持。

② 由于 ref 和 out 同属按引用传递，不能通过 ref 和 out 的不同实现重载，即不能定义两个完全一样的方法，仅有参数 ref 和 out 不同。

③ 不使用 ref 或 out 修饰的参数，不一定就是按值传递的。例如，数组、集合等都是引用类型，故不必使用 ref 修饰，也是按引用传递的。

3.5　构造方法

构造方法是类的一个特殊成员，会在类实例化对象时自动调用，为对象开辟内存空间，并对类中的成员进行初始化。

C#中的每个类至少有一个构造方法，当声明一个类时，如果没有定义构造方法，系统会自动添加一个默认的没有参数的构造方法，在其方法体中没有任何代码，即什么也不做；反之，如果声明类时定义了构造方法，系统将不会自动添加默认的构造方法。

在一个类中定义构造方法，必须满足以下 3 个条件：

（1）方法名与类名相同。

（2）在方法名的前面没有返回值类型的声明。

（3）在方法中不能使用 return 语句返回一个值。

【例 3-8】 在类中定义一个有参数的构造方法。

【示例代码：chapter03\Solution1\ConstructionMethod】

程序代码如下：

```
namespace ConstructionMethod
{
    class Program
    {
        static void Main(string[] args)
        {
            Person p = new Person("李雷");
            p.Speak();
            Console.ReadLine();
        }
    }
    public class Person
    {
        private string _name;
        public string Name
        {
            get{return _name;}
            set{_name = value;}
        }
        //定义有参数的构造方法
        public Person(string name)
        {
            _name = name;
        }
        public void Speak()
        {
            Console.WriteLine("我是" + _name);
        }
    }
}
```

程序运行结果如图 3-8 所示。

图 3-8 ConstructionMethod 项目程序运行结果

【分析】

在本实例中，在定义 Person 类时声明了一个有参数的构造方法，参数是一个 string 类型变量，因此，在主方法中实例化 Person 时，需要传入一个 string 类型的实参。

3.6 访问修饰符与 static 关键字

3.6.1 访问修饰符

访问修饰符用于限定外界对类和方法的访问权限。在 C#中，访问修饰符共有 4 种，分别是 public、protected、internal 和 private，使用这 4 种访问修饰符可以组合成 5 个可访问级别，访问级别从高到低描述如下。

（1）public：公有访问，最高访问级别，访问不受任何限制。

（2）protected：保护访问，只限于本类和子类访问，实例不能访问。

（3）internal：内部访问，只限于本项目内访问，其他不能访问。

（4）protected internal：内部保护访问，只限于本项目中的类或子类访问，其他不能访问。

（5）private：私有访问，最低访问级别，只限于在声明它们的类和结构中才可以访问，子类和实例都不能访问。

在使用访问修饰符定义命名空间、结构和类及其成员时，要注意以下几点：

（1）一个成员或类型只能有一个访问修饰符，使用 protected internal 组合时除外。

（2）命名空间上不允许使用访问修饰符，命名空间没有访问限制。

（3）如果未指定访问修饰符，则使用默认的可访问性，类的成员默认为 private。

（4）访问修饰符只是控制外部对内部成员的访问，类的内部对自己的访问不受其限制，即在类的内部可以访问所有的类成员。

3.6.2 static 关键字

static 关键字用于修饰类、字段、属性、方法以及构造方法等。被 static 修饰的成员称为静态成员，包括静态字段、静态属性、静态方法等；被 static 修饰的类称为静态类。

静态成员与非静态成员的不同在于：静态成员属于类，而不属于类的实例，因此，需要通过类而不通过类的实例来访问；而非静态成员总是与特定的实例（对象）相联系。在实际应用中，当类的成员引用或操作的信息是属于类而不属于类的实例时，就应该设置为静态成员。

1. 静态字段

静态字段不属于任何对象，只属于类，访问静态字段的语法格式如下：

类名.静态字段名

【例3-9】 定义一个静态字段,并访问该字段。
【示例代码:chapter03\Solution1\Static\ StaticField】
程序代码如下:

```
namespace Static
{
    class StaticField
    {
        static void Main(string[] args)
        {
            ComputerStudent stu1 = new ComputerStudent();
            stu1._name = "李雷";
            Console.WriteLine(stu1._name + "的专业是: " + ComputerStudent._specialtyName);
            Console.ReadLine();
        }
    }
    class ComputerStudent
    {
        public string _name;
        //定义静态字段
        public static string _specialtyName = "计算机科学与技术";
    }
}
```

程序运行结果如图 3-9 所示。

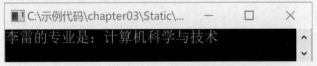

图 3-9 StaticField 类程序运行结果

【分析】

在本实例中,ComputerStudent 中声明了一个静态字段_specialtyName,在主方法中访问该字段时,直接使用 ComputerStudent._specialtyName 的形式即可,不能通过对象 stu1 进行访问。

2. 静态属性

静态属性可以读写静态字段的值,并保证静态字段值的合法性,访问静态属性的语法格式如下:

类名.静态属性名

【例 3-10】 定义一个静态属性,并访问该属性。

【示例代码：chapter03\Solution1\Static\ StaticProperty】
程序代码如下：

```
namespace Static
{
    class StaticProperty
    {
        static void Main(string[] args)
        {
            Student stu1 = new Student();
            stu1._name = "李雷";
            Console.WriteLine(stu1._name + "的专业是: " + Student.SpecialtyName);
            Console.ReadLine();
        }
    }
    class Student
    {
        public string _name;
        public static string _specialtyName = "计算机科学与技术";
        //定义静态属性
        public static string SpecialtyName
        {
            get
            {
                return _specialtyName;
            }
            set
            {
                _specialtyName = value;
            }
        }
    }
}
```

程序运行结果如图 3-10 所示。

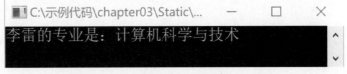

图 3-10　StaticProperty 类程序运行结果

【分析】
在本实例中，Student 中定义了静态属性 SpecialtyName，在主方法中直接通过类名.静态属性名，即 Student.SpecialtyName 的形式访问。

3. 静态方法

当希望不创建对象就可以访问某个方法时，可以将该方法定义成静态方法，访问静态方法的语法格式如下：

类名.静态方法名

【例 3-11】 定义一个静态方法，并访问该方法。

【示例代码：chapter03\Solution1\Static\StaticMethod】

程序代码如下：

```
namespace Static
{
    class StaticMethod
    {
        static void Main(string[] args)
        {
            ComStudent.Speak();
            Console.ReadLine();
        }
    }
    class ComStudent
    {
        public static void Speak()
        {
            Console.WriteLine("我的专业是：计算机科学与技术");
        }
    }
}
```

程序运行结果如图 3-11 所示。

图 3-11　StaticMethod 类程序运行结果

【分析】

在本实例中，ComStudent 中定义了一个静态的 Speak()方法，在主方法中直接通过类名访问该方法，即 ComStudent.Speak()。

4. 静态类

当类中的成员全部是静态成员时，就可以把这个类声明为静态类。静态类具有许多优点。例如，编译器能够自动执行检查，以确保不添加实例成员；静态类能够使程序的实现更简单、迅速，因为不必创建对象就能调用其方法。静态类具有以下特点：

（1）静态类仅包含静态成员。

（2）静态类不能被实例化。
（3）静态类是密封的。
（4）静态类不能包含实例构造方法。

【例 3-12】 定义一个静态类，并访问该类。

【示例代码：chapter03\Solution1\Static\StaticClass】

程序代码如下：

```
namespace Static
{
    class StaticClass
    {
        static void Main(string[] args)
        {
            CStudent.Speak();
            Console.ReadLine();
        }
    }
    public static class CStudent
    {
        public static string _specialtyName = "计算机科学与技术";
        public static void Speak()
        {
            Console.WriteLine("我的专业是：" + _specialtyName);
        }
    }
}
```

程序运行结果如图 3-12 所示。

图 3-12　StaticClass 类程序运行结果

【分析】

在本实例中，CStudent 是静态类，其中_specialtyName 字段和 Speak()方法都必须是静态的。在主方法中，静态类 CStudent 不能实例化，只能通过"类名.方法名"的形式，即 CStudent.Speak()进行访问。

3.7　面向对象的基本特征

面向对象的三个基本特征是封装、继承和多态。

封装是指将客观事物封装成抽象的类，并且类可以把自己的数据和方法只让可信的类或对象操作,对不可信的类进行信息隐藏。通常,一个类就是一个封装了数据以及操作这些数据的代码逻辑实体。

继承是指可以让某个类型的对象获得另一个类型对象的属性和方法。通过继承创建的新类称为"子类"或"派生类",被继承的类称为"基类""父类"或"超类"。继承的过程,就是从一般到特殊的过程。

多态是指一个类实例的相同方法在不同情形有不同的表现形式。多态机制使具有不同内部结构的对象可以共享相同的外部接口。这意味着,虽然针对不同对象的具体操作不同,但通过一个公共的类,这些操作可以采用相同的方式予以调用。

3.7.1 封装

C#中可以使用类达到数据封装的效果,这样就可以使数据与方法封装成单一元素,以便通过方法存取数据。

【例3-13】 封装。

【示例代码:chapter03\Solution1\Feature\MyClass1】

MyClass1 类的程序代码如下:

```
namespace Feature
{
    class MyClass1
    {
        public int X { get; set; }
        public int Y { get; set; }
        public int Add()
        {
            return X + Y;
        }
        public virtual int Subtract()
        {
            return X - Y;
        }
    }
}
```

【示例代码:chapter03\Solution1\Feature\Encapsulation】

Encapsulation 类的程序代码如下:

```
namespace Feature
{
    class Encapsulation
    {
        static void Main(string[] args)
        {
            MyClass1 myclass = new MyClass1();
            myclass.X = 48;
```

```
            myclass.Y = 9;
            Console.WriteLine("48 + 9 = "+myclass.Add());
            Console.ReadLine();
        }
    }
}
```

程序运行结果如图 3-13 所示。

图 3-13　Encapsulation 类程序运行结果

【分析】

本实例中,首先定义了一个类 MyClass1 来封装加数和被加数的属性,并通过 Add() 方法返回两个数的和,通过虚方法 Subtract() 返回两个数的差。在测试类 Encapsulation 中, 通过调用 MyClass1 对象的 Add() 方法实现两个整数求和。

3.7.2　继承

C#中提供了类的继承机制,但只支持单继承,而不支持多重继承,即在 C#中一次只允许继承一个类,不能同时继承多个类。利用类的继承机制,可以通过增加、修改或替换类中的方法对这个类进行扩充,以适应不同的应用需求。继承使子类可以从父类自动地获得父类所具备的特性,故可以极大地节省代码,提高代码的可重用性。

实现继承的语法格式如下:

```
[访问修饰符]class 类名:基类名
{
    类成员;
}
```

【例 3-14】 继承。

【示例代码:chapter03\Solution1\Feature\Inheritance】

程序代码如下:

```
namespace Feature
{
    class Inheritance
    {
        static void Main(string[] args)
        {
            MyClass2 myclass2 = new MyClass2();
            myclass2.X = 48;
            myclass2.Y = 9;
```

```
            myclass2.Z = 11;
            Console.WriteLine("48 + 9 = "+myclass2.Add());
            Console.WriteLine("48 + 9 + 11 = "+myclass2.Add2());
            Console.ReadLine();
        }
        class MyClass2 : MyClass1
        {
            public int Z { get; set; }
            public int Add2()
            {
                return X + Y + Z;
            }
        }
    }
}
```

程序运行结果如图 3-14 所示。

图 3-14 Inheritance 类程序运行结果

【分析】
本实例中定义了 MyClass2 类，该类继承于 MyClass1 类（例 3-13 中已声明），并扩展其成员方法 Add2()求 3 个整数的和。在测试类 Inheritance 中，通过 MyClass2 类的对象调用 MyClass1 类中的 Add()和 Add2()方法，实现了整数求和。

注意：
在 C#中，类的继承遵循以下原则。
（1）派生类只能从一个类中继承，即单继承。
（2）派生类自然继承基类的成员，但不能继承基类的构造方法。
（3）类的继承可以传递。例如，假设类 C 继承于类 B，类 B 又继承类 A，那么 C 类即具有类 B 和类 A 的成员，可以认为类 A 是类 C 的祖先类。

3.7.3 多态

在 C#中，类的多态性是通过在子类（派生类）中重写基类的虚方法或方法成员实现的。若一个方法在定义中含有 virtual 修饰符，则该方法称为虚方法。虚方法的实现可以由派生类取代，取代所继承的虚方法的实现过程称为重写（override）。

【例3-15】 多态。

【示例代码：chapter03\Solution1\Feature\Polymorphism】

程序代码如下：

```
namespace Feature
{
    class Polymorphism
    {
        static void Main(string[] args)
        {
            MyClass1 myclass1 = new MyClass1();
            myclass1.X = 3;
            myclass1.Y = 5;
            Console.WriteLine("两数差是：" + myclass1.Subtract());
            MyClass3 myclass3 = new MyClass3();
            Console.WriteLine("两数差是：" + myclass3.Subtract());
            Console.ReadLine();
        }
    }
    class MyClass3 : MyClass1
    {
        public override int Subtract()
        {
            int x = 15;
            int y = 7;
            return x - y;
        }
    }
}
```

程序运行结果如图 3-15 所示。

图 3-15　Polymorphism 类程序运行结果

【分析】

本实例中定义了一个 MyClass3 类，该类继承于 MyClass1（例 3-13 中已声明），在 MyClass3 类中重写虚方法 Subtract()。在测试类 Polymorphism 中，分别使用 MyClass1 和 MyClass3 类的对象调用 Subtract()方法，实现两整数求差。

习题3

第4章 WinForm 基础

学习目标：
- 掌握窗体的创建方法及其程序设计过程；
- 理解常用 WinForm 控件的功能，掌握其属性和事件编程；
- 了解 SDI 与 MDI 的区别及其程序设计方法。

4.1 WinForm 简介

WinForm 是 .NET 开发平台提供的用于开发 Windows 应用程序的一个开发环境，命名空间 System.Windows.Forms 中包含用于创建 Windows 应用程序用户界面所需要的类。

4.1.1 WinForm 程序的新建

右击"解决方案"，选择"添加"→"新建项目"选项，打开"添加新项目"对话框。在"添加新项目"对话框中，选择"Windows 窗体应用（.NET Framework）"选项，单击"下一步"按钮，进入"配置新项目"对话框，如图 4-1 所示。自定义项目名称并选择保存位置（这里命名为 WindowsFormsApp1），然后单击"创建"按钮创建一个 WinForm 程序。

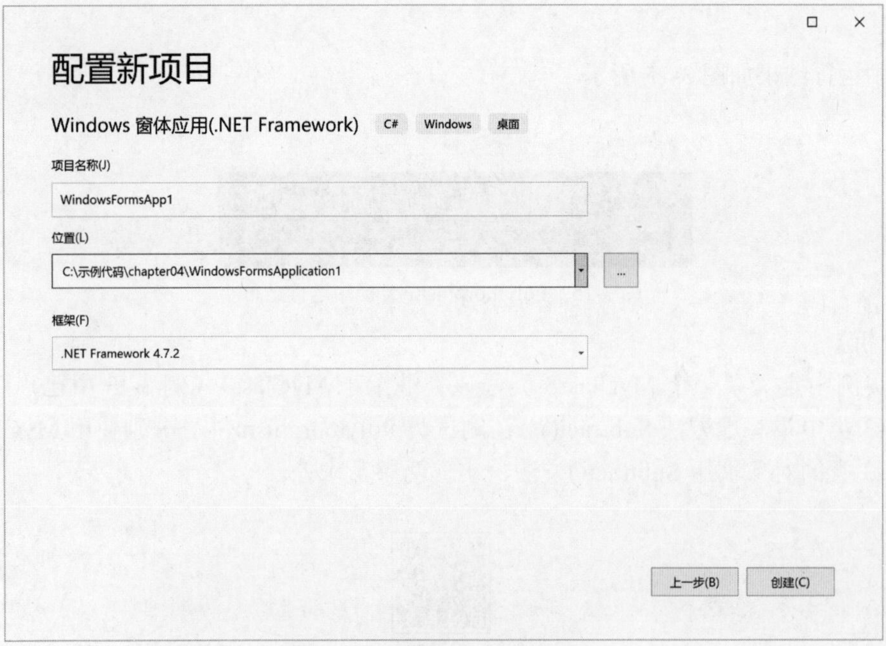

图 4-1 创建 Windows 窗体项目

4.1.2　WinForm 程序的文件结构

WinForm 程序中包含多种不同类型的文件。新建的 WindowsFormsApplication1 项目的文件结构如图 4-2 所示。

从图 4-2 中可以看出，WinForm 程序的文件结构包含 5 部分，分别是 Properties、引用、App.config、Form1.cs 和 Program.cs。其中，Properties 用来设置项目的属性；引用用来设置对其他程序集的引用；App.config 用来存储项目的配置信息；Form1.cs 用来设置窗体界面以及编写逻辑代码；Program.cs 用来设置项目运行时的主窗体。Form1.cs 与 Program.cs 专门用来实现窗体界面的设计和运行。

1．Form1.cs

从图 4-2 中可以看出，Form1.cs 由 Form1 与 Form1.Designer.cs 两部分组成，其中 Form1 文件又包括 Form1.cs[设计]界面和 Form1.cs 逻辑代码界面两部分。

（1）Form1.cs[设计]界面。

Form1 是 Form1.cs[设计]界面中系统初始化的窗体。如图 4-3 所示，在默认情况下，该窗体中没有任何控件。

图 4-2　WinForm 程序的文件结构　　　　图 4-3　Form1.cs 设计界面

（2）Form1.cs 逻辑代码界面。

右击窗体 Form1，选择"查看代码"，就可以进入 Form1.cs 逻辑代码界面，如图 4-4 所示。

注意：Form 窗体有两种模式：设计模式和编码模式。在设计模式下可以设计界面，并修改控件的属性等；在编码模式下可以编写代码。

（3）Form1.Designer.cs。

Form1.Designer.cs 文件用于在窗体类中自动生成控件的初始化代码，Form1.Designer.cs 文件代码与 Form1.cs 逻辑代码界面的代码组合到一起，就是 Form1 窗体的类文件。

```csharp
using System;
using System.Collections.Generic;
using System.ComponentModel;
using System.Data;
using System.Drawing;
using System.Linq;
using System.Text;
using System.Threading.Tasks;
using System.Windows.Forms;

namespace WindowsFormsApp1
{
    public partial class Form1 : Form
    {
        public Form1()
        {
            InitializeComponent();
        }
    }
}
```

图 4-4　Form1.cs 逻辑代码界面

2．Program.cs

每一种可执行程序都有主入口，默认情况下，Program.cs 文件就是 WinForm 程序的主入口，如图 4-5 所示，该程序运行的主窗体是 Form1，可以根据程序需求修改为其他窗体。

```csharp
using System;
using System.Collections.Generic;
using System.Linq;
using System.Threading.Tasks;
using System.Windows.Forms;

namespace WindowsFormsApp1
{
    static class Program
    {
        /// <summary>
        /// 应用程序的主入口点。
        /// </summary>
        [STAThread]
        static void Main()
        {
            Application.EnableVisualStyles();
            Application.SetCompatibleTextRenderingDefault(false);
            Application.Run(new Form1());
        }
    }
}
```

图 4-5　Program.cs 文件

4.1.3 窗体与控件

Windows 窗体和控件是开发 C# 应用程序的基础，每个 Windows 窗体和控件都是一个对象。

1．窗体

窗体是一个可以用来为用户提供信息以及接收其输入的窗口。窗体是其他对象的载体或容器，在窗体上可以直接创建应用程序，可以放置应用程序所需的控件以及图形、图像，并可以改变其大小，移动其位置。每个窗体对应于应用程序运行的一个窗口。

在 WinForm 项目上右击，在弹出的快捷菜单上选择"添加"→"窗体（Windows 窗体）"命令。在"添加新项"对话框中，如图 4-6 所示，选择"窗体（Windows 窗体）"选项，自定义名称（这里命名为 Form2.cs），然后单击"添加"按钮，就可以添加一个新的窗体。删除窗体时在要删除的窗体名称上右击，在弹出的快捷菜单上选择"删除"命令，即可删除窗体。

图 4-6 新建 Windows 窗体

2．控件

控件是能够提供用户界面接口功能的组件。控件可以通过属性设置控制其显示效果，可以对相应的事件做出反应，实现控制或交互功能。

选择"视图"→"工具箱"命令，会显示出"工具箱"窗口，如图 4-7 所示。为窗体添加一般控件，直接从工具箱中拖曳控件到窗体设计界面即可。

图 4-7 工具箱

如果为窗体添加第三方控件,可以在常用控件中右击,在弹出的快捷菜单中选择"选择项"命令,在打开的"选择工具箱项"对话框中,选择要添加的第三方控件后,单击"确定"按钮,如图 4-8 所示。

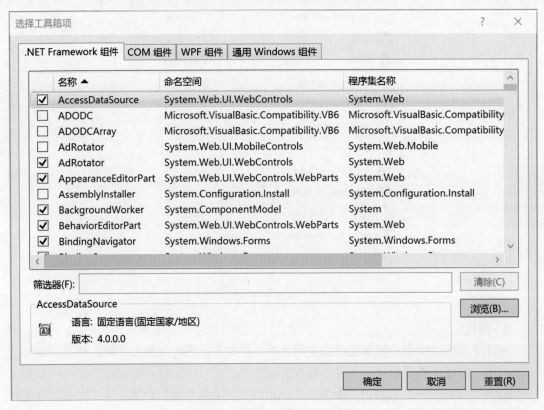

图 4-8 "选择工具箱项"对话框

4.1.4 属性与事件

1. 属性

窗体和控件都有许多属性,窗体与控件的常用属性如表 4-1 所示。

表 4-1 窗体和控件的常用属性

属　　性	说　　明
Name	指示代码中用来标识该对象的名称
Text	窗体标题或与控件关联的文本
BackColor	窗体或控件的背景色
Enabled	指示是否启用该控件
Size	窗体或控件的大小
Font	指示控件中文件的字体

设置属性有两种方法,一是通过属性窗口设置,该方法主要适用于在设计时设置窗体属性;二是通过代码设置,该方法主要适用于在编码时设置窗体属性。

(1)通过"属性"窗口设置属性。

单击要设置属性的窗体或控件,在"属性"窗口中找到要设置的属性,直接修改属性值。例如,设置窗体 Form1 的 Text 属性为"主窗体",如图 4-9 所示。

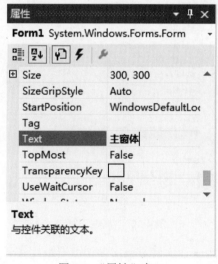

图 4-9 "属性"窗口

(2)通过代码设置属性。

设置属性的代码格式如下:

窗体唯一标识.属性名 = 属性值

例如,设置窗体 Form1 的标题为"主窗体",程序代码如下:

```
Form1.Text = "主窗体";
```

2. 事件

事件指的是可能发生在对象上的，能够被该对象识别同时通过代码响应或处理的行为。事件可由用户操作、程序代码或系统触发。事件处理程序是绑定到事件的方法。

每个窗体和控件都公开一组预定义的事件，可根据这些事件进行编程。窗体与控件的事件如表 4-2 所示。

表 4-2 窗体和控件的常用事件

事件	说明
Load	窗体加载时被触发
Click	单击控件时触发
DoubleClick	双击控件时触发
FormClosed	窗体关闭后触发
BackColorChanged	控件的背景色值更改时触发

如果为窗体或控件添加事件，首先选中要添加事件的窗体或控件，单击"属性"窗口的"⚡"图标，将显示所有事件，如图 4-10 所示。然后选择要添加的事件，在其后面的空格中双击，会进入窗体代码界面，并自动生成相应的事件处理程序格式。

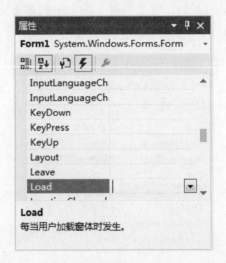

图 4-10 "属性"窗口

【例 4-1】 设计一个播放器，当单击播放按钮时，播放指定视频文件（在 chapter04 文件夹下的 clock.avi 文件）。

【操作步骤】

（1）启动 VS，新建一个 Windows 窗体应用程序 MediaPlay。

（2）为工具箱添加第三方控件 Windows Media Player。

（3）双击 Form1.cs，切换到设计视图，从工具栏中拖曳一个 Windows Media Player 控件、一个 Button 控件到窗体设计区，并调整 Windows Media Player 控件大小进行布局。

（4）在窗体设计区中右击窗体 Form1 和每个控件，设置窗体和控件的相关属性。表 4-3 列出了窗体及控件属性。

表 4-3　窗体及控件属性设置

窗体和控件	属　　性	属　性　值
Form1	Text	播放器
axWindowsMediaPlayer1	Name	media
button1	Name	btnPlay
	Text	播放

（5）双击"播放"按钮，为其添加单击事件处理程序，程序代码如下。

```
private void btnPlay_Click(object sender, EventArgs e)
{
    media.URL = @" C:\示例代码\chapter04\clock.avi";
}
```

上述代码的功能是，当单击"播放"按钮时，系统将播放 C:\示例代码\chapter04 目录下的 clock.avi 文件。

注意：事件处理程序通常有两个参数：第一个参数引用触发事件的控件；第二个参数封装了事件的一些状态信息。不同事件的处理过程，其第二个参数的类型也不同，相应的，其内部封装的状态信息也不同。

（6）在解决方案资源管理器中右击 MediaPlay 项目，将其设为启动项目。

（7）编译并运行，运行结果如图 4-11 所示。

图 4-11　"播放器"界面

4.2　WinForm 常用控件

控件是界面组件，是带有可视化表示形式的组件（组件包含控件和不能可视化的组件），是包含在窗体内的对象。控件的主要功能就是实现输入和输出，每种类型的控件都具有其特有的属性和事件。

4.2.1 文本类控件

文本类控件包括标签控件（Label）、按钮控件（Button）、文本框控件（TextBox）以及有格式文本框控件（RichTextBox）等。

1. Label 控件

Label，即标签控件，是由 System.Windows.Forms.Label 类提供的，主要用于显示文本。Label 控件的常用属性和事件如表 4-4 所示。

表 4-4　Label 控件的常用属性和事件

类别	名称	用途
属性	Name	指示代码中用来标识该对象的名称
	Text	显示的文本
	Visible	确定该控件是可见的还是隐藏的
	Font	显示控件中文本的字体
事件	Click	单击事件

2. Button 控件

Button，即按钮控件，是由 System.Windows.Forms.Button 类提供的，主要用于接收用户对鼠标的操作，完成用户与应用程序之间的交互。Button 控件支持的操作包括鼠标的单击、双击以及键盘的 Enter 键操作。在设计时，先添加 Button 控件到窗体设计区，然后双击它，即可编写 Click 事件处理程序。在执行程序时，只要单击该按钮，就会执行 Click 事件中的代码。Button 控件的常用属性和常用事件如表 4-5 所示。

表 4-5　Button 控件的常用属性和事件

类别	名称	用途
属性	Name	指示代码中用来标识该对象的名称
	Text	按钮上显示的文本
	Size	控件的大小（以像素为单位）
	Image	设置控件的图像
事件	Click	单击事件

3. TextBox 控件

TextBox，即文本框控件，是由 System.Windows.Forms.TextBox 类提供的，主要用于在应用程序中接收用户输入的文字。它允许用户在其中输入任何字符，当然也可以指定输入某种类型的字符，如只允许输入数字。文本框支持 3 种输入模式，包括单行文本模式、多行文本模式和密码输入模式。默认情况下，文本框工作在单行文本模式，最多可输入 2048 个字符。设置文本框的 Multiline 属性为 True，表示指定文本框为多行文本模式，此时最多可输入 32KB 的文本。如果指定了 UseSystemPasswordChar 属性为 True，则文本框为密码输入模式，此时无论用户输入什么文本，系统只显示密码字符。用户的所有输入保存在 Text 属性中，在程序中引用 Text 属性，即可获得用户输入的文本。TextBox 控件的常用属性和事件如表 4-6 所示。

表 4-6 TextBox 控件的常用属性和事件

类别	名称	用途
属性	Name	指示代码中用来标识该对象的名称
	Text	显示的文本
	ScrollBars	指定对于多行编辑控件,将为此控件显示哪些滚动条
	PasswordChar	设置密码字符串
	UseSystemPasswordChar	指示编辑控件中的文本是否以默认的密码字符显示
	ReadOnly	指示文框中的文本是否为只读
	Multiline	控制编辑控件的文本是否能跨越多行
事件	Enter	进入控件时引发的事件
	Leave	失去输入焦点时引发的事件
	TextChanged	在控件上更改 Text 属性的值时引发的事件

【例 4-2】 设计一个简单的用户登录界面,当输入用户名和密码正确时,提示登录成功,否则登录失败。

【操作步骤】

(1)启动 VS,新建一个 Windows 窗体应用程序 Controls。

(2)双击 Form1.cs,切换到设计视图,从工具栏中拖曳 2 个 Label 控件,2 个 TextBox 控件和 2 个 Button 控件到窗体设计区,并根据图 4-12 调整控件大小和布局。

(3)在窗体设计区中右击窗体 Form1 和每一个控件,设置窗体和控件的相关属性。表 4-7 列出了窗体和控件的属性设置。

表 4-7 窗体和控件的属性设置

窗体和控件	属性	属性值
Form1	Text	登录
label1	Text	用户名:
label2	Text	密码:
textBox1	Name	txtName
textBox2	Name	txtPassword
	UseSystemPasswordChar	True
button1	Name	btnLogin
	Text	登录
button2	Name	btnReset
	Text	重置

注意:将文本框设置成密码框有两种方法:第一种是将 PasswordChar 属性值设为"*"或其他密码符号;第二种是直接将 UseSystemPasswordChar 属性设置为 True。

(4)双击"登录"按钮,为其添加单击事件处理程序,程序代码如下。

```
private void btnLogin_Click(object sender, EventArgs e)
{
    if (txtName .Text == "user" && txtPassword .Text == "123")
    {
```

```
            MessageBox.Show("登录成功!");
        }
        else
        {
            MessageBox.Show("用户名或密码错误!");
        }
    }
```

上述代码的功能是，当在 txtName（"用户名"文本框）中输入 user，并在 txtPassword（"密码"文本框）中输入 123 之后，单击 btnLogin（"登录"按钮），系统将弹出消息对话框提示"登录成功!"，否则，将提示"用户名或密码错误!"。

（5）双击"重置"按钮，为其添加单击事件处理程序，程序代码如下。

```
private void btnReset_Click(object sender, EventArgs e)
{
    //遍历窗体中所有控件
    foreach (Control item in this.Controls)
    {
        if (item is TextBox)
        {
            item.Text = "";
        }
    }
}
```

以上代码的功能是，使用 foreach 循环遍历窗体中所有控件，并设置每个控件的 Text 属性值为空，即实现了清除输入内容的功能。

（6）在解决方案资源管理器中右击 Controls 项目，将其设为启动项目。

（7）编译并运行，运行结果如图 4-12 所示。

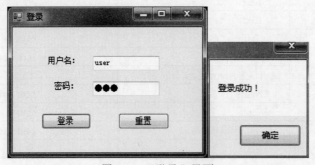

图 4-12 "登录"界面

4．RichTextBox 控件

RichTextBox，即有格式文本框控件，是由 System.Windows.Forms.RichTextBox 类提供的，主要用于打开和保存 RTF 文件或普通的 ASCII 文本文件，并且显示、输入和操作带有格式的文本。RichTextBox 的功能与 TextBox 类似，但也有一些不同的地方，TextBox 控件常用于输入较短的文本字符，而 RichTextBox 多用于显示和输入格式化的文本。RichTextBox 控件使用标准的格式化文本，称为富文本格式（Rich Text Format，RTF），可以显示字体、

颜色和链接，从文件加载文本和加载嵌入的图像，以及查找指定的字符，因此 RichTextBox 常常被称为增强的文本框。RichTextBox 控件可以显示滚动条，这一点与 TextBox 控件相同。与 TextBox 控件不同的是，RichTextBox 控件的默认设置是水平和垂直滚动条均根据需要显示，并且拥有更多的滚动条设置。RichTextBox 控件的常用属性和事件如表 4-8 所示。

表 4-8 RichTextBox 控件的常用属性和事件

类 别	名 称	用 途
属性	Name	指示代码中用来标识该对象的名称
	Lines	多行编辑中的文本行，作为字符串值的数组
	Size	控件的大小（以像素为单位）
	Location	控件左上角相对于其容器左上角的坐标
事件	TextChanged	在控件上更改 Text 属性的值时引发的事件

【例 4-3】 设计一个简历编辑器，实现对 jianli.rtf 文件（在 Controls 项目 bin\Debug 文件夹中，新建一个 jianli.rtf 文件）的各种操作。

【操作步骤】

（1）启动 VS，在 Windows 窗体应用程序 Controls 项目中新建一个窗体 Form2。

（2）双击 Form2.cs，切换到设计视图，从工具栏中拖曳 1 个 RichTextBox 控件和 8 个 Button 控件到窗体设计区，并根据图 4-13 调整控件大小和布局。

（3）在窗体设计区中右击窗体 Form2 和每一个控件，设置窗体和控件的相关属性。表 4-9 列出了窗体和控件的属性设置。

表 4-9 窗体和控件的属性设置

窗体和控件	属 性	属 性 值
Form2	Text	简历编辑器
RichTextBox1	Name	richTextBox1
Button1	Name	btnLoad
	Text	加载
Button2	Name	btnSave
	Text	保存
Button3	Name	btnCopy
	Text	复制
Button4	Name	btnCut
	Text	剪切
Button5	Name	btnPaste
	Text	粘贴
Button6	Name	btnBackColor
	Text	背景色
Button7	Name	btnColor
	Text	颜色
Button8	Name	btnFont
	Text	字体

（4）定义全局变量 file，用来存储所要编辑的简历文件名称，程序代码如下。

```
string file = "jianli.rtf";
```

（5）双击"加载"按钮，为其添加单击事件处理程序，程序代码如下：

```
private void btnLoad_Click(object sender, EventArgs e)
{
    richTextBox1.LoadFile(file);
}
```

上述代码的功能是，当单击"加载"按钮时，则在 RichTextBox 控件中打开文件 jianli.rtf。

（6）双击"保存"按钮，为其添加单击事件处理程序，程序代码如下：

```
private void btnSave_Click(object sender, EventArgs e)
{
    richTextBox1.SaveFile(file);
}
```

以上代码的功能是，当对简历文件编辑完成之后，单击"保存"按钮，将编辑后的文件内容保存到文件 jianli.rtf 中。

（7）分别双击"复制""剪切""粘贴"按钮，为其添加单击事件处理程序，程序代码如下：

```
private void btnCopy_Click(object sender, EventArgs e)
{
    richTextBox1.Copy();
}
private void btnCut_Click(object sender, EventArgs e)
{
    richTextBox1.Cut();
}
private void btnPaste_Click(object sender, EventArgs e)
{
    richTextBox1.Paste();
}
```

以上代码的功能是，在 RichTextBox 控件中，可以实现对 RichTextBox 控件中选中文本内容的复制、剪切和粘贴操作。

（8）分别双击"背景色""颜色""字体"按钮，为其添加单击事件处理程序，程序代码如下：

```
private void btnBackColor_Click(object sender, EventArgs e)
{
    //背景色更改为黄色
    richTextBox1.SelectionBackColor = Color.Yellow;
}
private void btnColor_Click(object sender, EventArgs e)
{
    //文字颜色更改为蓝色
    richTextBox1.SelectionColor = Color.Blue;
}
```

```
private void btnFont_Click(object sender, EventArgs e)
{
    //字体更改为20号黑体
    Font font = new Font("黑体", 20);
    richTextBox1.SelectionFont = font;
}
```

以上代码的功能是，通过单击相应按钮，实现对 RichTextBox 中选中内容的格式设置。例如，将背景色设置为黄色，将文字颜色设置为蓝色，将字体设置为 20 号黑体。

（9）在解决方案资源管理器 Controls 项目中双击 Program.cs 文件，将 Main()方法中的最后一行代码修改如下：

```
Application.Run(new Form2());
```

以上代码的功能是，Controls 项目运行时启动 Form2 窗体。

（10）编译并运行，运行结果如图 4-13 所示。

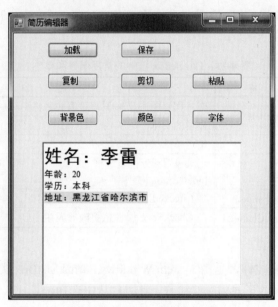

图 4-13 "简历编辑器"界面

4.2.2 选择类控件

在使用文本框构建用户的输入界面时，虽然可以检查或验证用户输入的有效性，但仍然不能完全确保用户的输入是系统所期望的数据。为此，必须设计只能通过选择即可完成数据输入的操作界面。.NET Framework 为 Windows 窗体提供了丰富的选择类控制，选择类控件包括单选按钮控件（RadioButton）、复选框控件（CheckBox）、列表框控件（ListBox）以及下拉框控件（ComboBox）等。

1．RadioButton 控件

RadioButton，即单选按钮控件，是由 System.Windows.Forms.RadioButton 类提供的，主要用于将一个或多个选项列出，让用户从中选择一项。RadioButton 控件的常用属性和事件如表 4-10 所示。

表 4-10 RadioButton 控件的常用属性和事件

类别	名称	用途
属性	Name	指示代码中用来标识该对象的名称
	Text	显示的文本
	Checked	指示单选按钮是否选中
	AutoSize	指定控件是否自动调整自身的大小以适应其内容的大小
事件	CheckedChanged	当 Checked 属性更改值时发生

2．CheckBox 控件

CheckBox，即复选框控件，是由 System.Windows.Forms.CheckBox 类提供的，主要用于将一个或多个选项列出，让用户从中选择一项或多项。当某一个选项被选中后，其左边的小方框会显示有一个勾。CheckBox 控件和 RadioButton 控件具有一个相似的功能，都允许用户从选项列表中进行选择。CheckBox 控件允许用户选择多个选项，而 RadioButton 控件用来构造相互排斥的选项组，并且只允许用户从中选择一个。CheckBox 控件的常用属性和事件如表 4-11 所示。

表 4-11 CheckBox 控件的常用属性和事件

类别	名称	用途
属性	Name	指示代码中用来标识该对象的名称
	Text	显示的文本
	Checked	指示单选按钮是否选中
	CheckState	用来设置或返回复选框的状态，有 3 种可能：Checked、Unchecked、Indeterminage
事件	CheckedChanged	当 Checked 属性值更改时发生
	CheckStateChanged	当 CheckState 属性值更改时发生

3．ListBox 控件

ListBox，即列表框控件，是由 System.Windows.Forms.ListBox 类提供的，主要用于将一个集合数据以列表框的形式显示给用户，供用户从中选择一项或多项。ListBox 控件有两种工作模式：单选模式和多选模式。当工作在单选模式时，列表框与单选按钮的功能相同；当工作在多选模式时，则与复选框功能相同。所不同的是，需要多个 RadionButton 控件或 CheckBox 控件才能构造一个选项组，而使用 ListBox 控件，则只需要一个就可以产生一个选项列表。ListBox 控件的常用属性和事件如表 4-12 所示。

表 4-12 ListBox 控件的常用属性和事件

类别	名称	用途
属性	Name	指示代码中用来标识该对象的名称
	Enabled	指示是否启用该控件
	Items	列表框中的项目集合
	ItemHeight	列表框中项的高度
	SelectionMode	指示列表框将是单项、多项还是不可选择
事件	SelectedIndexChanged	当 SelectedIndex 属性值更改时发生

注意：当 ListBox 的 SelectionMode 的属性值为 MultiSimple 或 MultiExtended 时，SelectedIndex 返回的是选中的最小索引，SelectedItem 返回的是选中的索引值最小的选项值。Items 本身也包含很多属性，如 Count 属性指示 Items 包含项的个数等。

4．ComboBox 控件

ComboBox，即下拉框控件，是由 System.Windows.Forms.ComboBox 类提供的，主要用于将一个集合数据以下拉列表框显示给用户，供用户从中选择一项。ComboBox 控件的默认行为是显示一个可编辑文本框，该文本框具有一个隐藏的下拉列表框。ComboBox 控件只支持单选，可替代 RadioButton 选项组。ComboBox 控件的常用属性和事件如表 4-13 所示。

表 4-13　ComboBox 控件的常用属性和事件

类　别	名　称	用　途
属性	Name	指示代码中用来标识该对象的名称
	Items	列表框中的项目集合
事件	SelectedIndexChanged	当 SelectedIndex 属性值更改时发生

【例 4-4】　设计一个注册界面。分别定义 RadioButton 控件显示性别，ComboBox 控件显示城市；CheckBox 控件显示技术方向；ListBox 控件显示兴趣爱好。当用户单击"注册"按钮时，如果填写的信息完整，提示"注册成功！"；否则，提示"请填写全部信息！"。

【操作步骤】

（1）启动 VS，在 Windows 窗体应用程序 Controls 项目中新建一个窗体 Form3。

（2）双击 Form3.cs，切换到设计视图，从工具栏中拖曳 6 个 Label 控件、2 个 TextBox 控件、2 个 RadioButton 控件、1 个 ComboBox 控件、3 个 CheckBox 控件、1 个 ListBox 控件和 1 个 Button 控件到窗体设计区，并根据图 4-14 调整控件大小和布局，并完成所有 Label 控件 Text 属性设置。

（3）在窗体设计区中右击窗体 Form3 和每一个控件，设置窗体和控件的相关属性。表 4-14 列出了除 Label 以外的窗体和控件的属性设置。

表 4-14　窗体和控件的属性设置

窗体和控件	属　性	属　性　值
Form3	Text	注册
textBox1	Name	txtName
textBox2	Name	txtPassword
	UseSystemPasswordChar	True
radioButton1	Name	rdMale
	Text	男
	Checked	True
radioButton2	Name	rdFemale
	Text	女
comboBox1	Name	city
	Items	北京/上海/哈尔滨/大连/深圳/广州
checkBox1	Name	net
	Text	.NET
checkBox2	Name	c
	Text	C
checkBox3	Name	java
	Text	JAVA

窗体和控件	属　性	属　性　值
listBox1	Name Items	interest 旅行/看书/运动
button1	Name Text	btnRegister 注册

(4) 定义一个名为 IsNullOrEmpty 的方法，用于判断用户是否已输入全部注册信息，程序代码如下：

```
public bool IsNullOrEmpty(Control.ControlCollection controls)
{
    bool flag = false;
    foreach (Control item in controls)
    {
        //判断当前控件内容是否为空
        if (string.IsNullOrEmpty(item.Text))
        {
            flag = true;
        }
    }
    return flag;
}
```

以上代码的功能是，判断窗体中所有控件内容是否为空，如果某个控件内容为空，则返回 true；否则，返回 false。

(5) 双击"注册"按钮，为其添加单击事件处理程序，程序代码如下：

```
private void btnRegister_Click(object sender, EventArgs e)
{
    bool flag = IsNullOrEmpty(this.Controls);
    if (flag == true)
    {
        MessageBox.Show("请填写全部信息！");
    }
    else
    {
        MessageBox.Show("注册成功！");
    }
}
```

以上代码的功能是，调用 IsNullOrEmpty 方法，实参为窗体中所有控件集合，方法返回值如果为 true，表示窗体中某控件 Text 属性值为空，即用户填写注册信息不全，则提示用户"请填写全部信息！"；否则，提示用户"注册成功！"。

(6) 在解决方案资源管理器 Controls 项目中双击 Program.cs 文件，将 Main()方法中的最后一行代码修改如下。

```
Application.Run(new Form3());
```

以上代码的功能是，Controls 项目运行时启动 Form3 窗体。

（7）编译并运行，运行结果如图 4-14 所示。

图 4-14 "注册"界面

4.2.3 分组类控件

分组类控件包括面板控件（Panel）、分组框控件（GroupBox）以及选项卡控件（TabControl）等。

1. Panel 控件

Panel，即面板控件，是由 System.Windows.Forms.Panel 类提供的，主要用于将其他控件组合在一起放在一个面板上，使这些控件更容易管理。Panel 控件的常用属性和事件如表 4-15 所示。

表 4-15 Panel 控件的常用属性和事件

类 别	名 称	用 途
属性	Name	指示代码中用来标识该对象的名称
	Size	控件的大小（以像素为单位）
	Location	控件左上角对于其容器左上角的坐标
	AutoScroll	指示当控件内容大于它的可见区域时是否自动显示滚动条
	BorderStyle	指示面板是否有边框
事件	Paint	在控件需要重新绘制时发生

【例 4-5】使用 Panel 控件设计一个显示列表，当单击"增加"按钮时，列表框中增加一项"子项"，当列表框大小超出可见区域时将自动显示滚动条。

【操作步骤】

（1）启动 VS，在 Windows 窗体应用程序 Controls 项目中新建一个窗体 Form4。

（2）双击 Form4.cs，切换到设计视图，从工具栏中拖曳 1 个 Panel 控件、1 个 ListBox 控件和 1 个 Button 控件到窗体设计区，并根据图 4-15 调整控件大小和布局。

(3) 在窗体设计区中右击窗体 Form4 和每一个控件，设置窗体和控件的相关属性。表 4-16 列出了窗体和控件的属性设置。

表 4-16 窗体和控件的属性设置

窗体和控件	属　　性	属　性　值
Form4	Text	面板
panel1	Name	panel
	AutoScroll	True
	BorderStyle	Fixed3D
button1	Name	btnAdd
	Text	增加
listBox1	Name	lbxText

(4) 双击"增加"按钮，为其添加单击事件处理程序，程序代码如下：

```
private void btnAdd_Click(object sender, EventArgs e)
{
    lbxText.Items.Add("子项");        //增加项
    lbxText.Height += 20;             //高度增加20像素
}
```

以上代码的功能是，每单击"增加"按钮一次，列表框中会增加一个 Items 项，内容是"子项"，同时，列表框的高度增加 20 像素。

(5) 在解决方案资源管理器 Controls 项目中双击 Program.cs 文件，将 Main()方法中的最后一行代码修改如下：

```
Application.Run(new Form4());
```

以上代码的功能是，Controls 项目运行时启动 Form4 窗体。

(6) 编译并运行，运行结果如图 4-15 所示。

图 4-15 "面板"界面

2. GroupBox 控件

GroupBox，即分组框控件，是由 System.Windows.Forms.GroupBox 类提供的，主要用于为其他控件提供可识别的分组。在设计窗体时，通常按功能把窗体划分为若干区域，每个区域使用一个 GroupBox 控件表示。例如，把相关的各选项放入一个分组框，就可以为用户提供一个统一的外观或逻辑处理。GroupBox 控件的常用属性和事件如表 4-17 所示。

表 4-17 GroupBox 控件的常用属性和事件

类 别	名 称	用 途
属性	Name	指示代码中用来标识该对象的名称
	Text	显示的文本
事件	Enter	在控件成为该窗体的活动控件时发生

3. TabControl 控件

TabControl，即选项卡控件，是由 System.Windows.Forms.TabControl 类提供的，主要用于将相关的组件组合到一系列选项卡页面上。TabControl 控件管理 TabPages 集合。TabControl 控件的常用属性和事件如表 4-18 所示。

表 4-18 TabControl 控件的常用属性和事件

类 别	名 称	用 途
属性	Name	指示代码中用来标识该对象的名称
	TabPages	选项卡集合
	Text	选项卡上显示的文本
	Appearance	指示选项卡是绘制成按钮还是绘制成常规选项卡
事件	Click	单击事件

【例 4-6】 设计一个分组窗体，当单击"选项"组中的"选项一"或"选项二"按钮时，在"标签"组中将对应显示 TabControl 控件中的选项卡。

【操作步骤】

（1）启动 VS，在 Windows 窗体应用程序 Controls 项目中新建一个窗体 Form5。

（2）双击 Form5.cs，切换到设计视图，从工具栏中拖曳 2 个 GroupBox 控件、2 个 Button 控件、1 个 TabControl 控件到窗体设计区，并根据图 4-16 调整控件大小和布局。

（3）在窗体设计区中右击窗体 Form5 和每一个控件，设置窗体和控件的相关属性。表 4-19 列出了窗体和控件的属性设置。

表 4-19 窗体和控件的属性设置

窗体和控件	属 性	属 性 值
Form5	Text	分组
groupBox1	Text	选项
groupBox2	Text	标签
tabControl1	TabPages	tabPage1/tabPage2
tabPage1	Name	tabPage1
	Text	选项一
tabPage2	Name	tabPage2
	Text	选项二

窗体和控件	属　性	属　性　值
button1	Name Text	btnOpt1 选项一
button2	Name Text	btnOpt2 选项二

（4）双击"选项一"按钮，为其添加单击事件处理程序，程序代码如下：

```
private void btnOpt1_Click(object sender, EventArgs e)
    {
        tabControl1.SelectedIndex = 0;
    }
```

以上代码的功能是，单击"选项一"按钮，则显示 TabControl 控件中的选项卡 tabPage1，其索引为 0。

（5）双击"选项二"按钮，为其添加单击事件处理程序，程序代码如下：

```
private void btnOpt2_Click(object sender, EventArgs e)
    {
        tabControl1.SelectedIndex = 1;
    }
```

以上代码的功能是，单击"选项二"按钮，则显示 TabControl 控件中的选项卡 tabPage2，其索引为 1。

（6）在解决方案资源管理器 Controls 项目中双击 Program.cs 文件，将 Main()方法中的最后一行代码修改如下：

```
Application.Run(new Form5());
```

以上代码的功能是，Controls 项目运行时启动 Form5 窗体。

（7）编译并运行，运行结果如图 4-16 所示。

图 4-16 "分组"界面

4.2.4 其他控件

1. 图片框控件

PictureBox，即图片框控件，是由 System.Windows.Forms.PictureBox 类提供的，主要用于在应用程序中显示图片，支持 Bitmap、Gif、Jpg 等多种图片格式。PictureBox 控件的常用属性和事件如表 4-20 所示。PictureBox 控件的 Image 属性是一个 Image 类的值，Image 类的对象用来保存图形信息，可使用 FromFile()方法将一个指定位置的图形文件加载到 Image 对象中。

表 4-20 PictureBox 控件的常用属性和事件

类 别	名 称	用 途
属性	Name	指示代码中用来标识该对象的名称
	Image	图片路径
	Size	控件的大小（以像素为单位）
	SizeMode	控制图片框将如何处理图像位置和控件大小
事件	Click	单击事件

【例 4-7】 设计一个图片显示窗体，当单击"显示图片"按钮时，在图片框中将显示指定图像（图片路径 C:\示例代码\chapter04\Koala.jpg）。

【操作步骤】

（1）启动 VS，在 Windows 窗体应用程序 Controls 项目中新建一个窗体 Form6。

（2）双击 Form6.cs，切换到设计视图，从工具栏中拖曳 1 个 PictureBox 控件和 1 个 Button 控件到窗体设计区，并根据图 4-17 调整控件大小和布局。

（3）在窗体设计区中右击窗体 Form6 和每一个控件，设置窗体和控件的相关属性。表 4-21 列出了窗体和控件的属性设置。

表 4-21 窗体和控件的属性设置

窗体和控件	属 性	属 性 值
Form6	Text	图片
pictureBox1	SizeMode	StretchImage
button1	Name	btnShow
	Text	显示图片

（4）双击"显示图片"按钮，为其添加单击事件处理程序，程序代码如下：

```
private void btnShow_Click(object sender, EventArgs e)
    {
        pictureBox1.Image = Image.FromFile(@"C:\示例代码\chapter04\Koala.jpg");
    }
```

以上代码的功能是，单击"显示图片"按钮，则显示路径为 C:\示例代码\chapter04\Koala.jpg 的图片。

（5）在解决方案资源管理器 Controls 项目中双击 Program.cs 文件，将 Main()方法中的最后一行代码修改如下：

```
Application.Run(new Form6());
```

以上代码的功能是，Controls 项目运行时启动 Form6 窗体。

（6）编译并运行，运行结果如图 4-17 所示。

图 4-17 "图片"界面

2．菜单控件

MenuStrip，即下拉菜单控件，是由 System.Windows.Forms.MenuStrip 类提供的，是应用程序菜单结构的容器。MenuStrip 控件的常用属性和事件如表 4-22 所示。

表 4-22 MenuStrip 控件的常用属性和事件

类 别	名 称	用 途
属性	Name	指示代码中用来标识该对象的名称
	Text	显示的文本
	Items	显示项的集合
事件	ItemClicked	当单击项时发生

在工具箱中直接双击 MenuStrip 控件，即可在窗体的顶部建立一个菜单，此时窗体的底部会显示出所创建的菜单名称，默认名称为 menuStrip1。把鼠标移到"请在此处键入"处，将会显示一个三角形按钮，单击该按钮将弹出一个下拉列表框，其中包括 MenuItem、ComboBox 和 TextBox 三个选项，默认为 MenuItem，如图 4-18 所示。

在"请在此处键入"处单击，即可在该文本框中输入文本，即设置菜单项的标题内容，输入内容后，在该文本的下方和右侧均会出现类似的"请在此处键入"字样，此时，可在下方为当前菜单创建子菜单，在右侧可以创建同一级别的其他菜单。

在输入标题内容时，可以在标题内容的某个字母前加"&"。例如，"窗体（&F）"命令将具有一个快捷键 Alt+F，程序运行时，按快捷键 Alt+F 同样可以执行此菜单命令。为菜单添加快捷键，也可以通过 ShorcutKeys 来设置。例如，设置 ShorcutKeys 属性为 Fom1 菜单添加了 Ctrl+K 快捷键，如图 4-19 所示。

图 4-18　创建菜单

图 4-19　输入菜单

【例 4-8】　设计一个菜单窗体，当单击"窗体"菜单中窗体名时，将打开对应的窗体界面，当单击"退出"菜单时，将退出程序。

【操作步骤】

（1）启动 VS，在 Windows 窗体应用程序 Controls 项目中新建一个窗体 Form7。

（2）双击 Form7.cs，切换到设计视图，从工具栏中添加一个 MenuStrip 控件到窗体设计区。

（3）在窗体设计区中右击窗体 Form7 和每一个控件，设置窗体和控件的相关属性。表 4-23 列出了窗体和控件的属性设置。

表 4-23　窗体和控件的属性设置

窗体和控件	属　性	属　性　值
Form7	Text	菜单
menuStrip1：Items	Name Text	formToolStripMenuItem 窗体（&F）
formToolStripMenuItem 1：DropDownItems	Name Text	form1ToolStripMenuItem Form1（&K）
formToolStripMenuItem 1：DropDownItems	Name Text	form2ToolStripMenuItem Form2（&L）
menuStrip1：Items	Name Text	exitToolStripMenuItem 退出（&E）

（4）双击"窗体"菜单中的 Form1 子菜单，为其添加单击事件处理程序，程序代码如下：

```
private void form1ToolStripMenuItem_Click(object sender, EventArgs e)
{
    //实例化Form1
    Form1 f = new Form1();
    //调用Show方法，显示窗体Form1
    f.Show();
}
```

以上代码的功能是，当单击 Form1 菜单时，则打开 Controls 项目中的 Form1 窗体。

（5）双击"窗体"菜单中的 Form2 子菜单，为其添加单击事件处理程序，程序代码如下：

```
private void form2ToolStripMenuItem_Click(object sender, EventArgs e)
{
    Form2 f = new Form2();
    f.Show();
}
```

以上代码的功能是，当单击 Form2 菜单时，则打开 Controls 项目中的 Form2 窗体。

（6）双击"退出"菜单，为其添加单击事件处理程序，程序代码如下：

```
private void exitToolStripMenuItem_Click(object sender, EventArgs e)
{
    //退出程序
    Application.Exit();
}
```

以上代码的功能是，当单击"退出"菜单时，程序退出。

（7）在解决方案资源管理器 Controls 项目中双击 Program.cs 文件，将 Main()方法中的最后一行代码修改如下：

```
Application.Run(new Form7());
```

以上代码的功能是，Controls 项目运行时启动 Form7 窗体。

（8）编译并运行，单击"窗体"菜单中 Form1 子菜单，则打开例 4-2 的窗体 Form1，运行结果如图 4-20 所示。

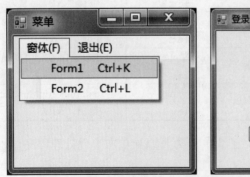 

图 4-20 "菜单"窗口及其运行结果

3．定时器控件

Timer，即定时器控件，是由 System.Windows.Forms.Timer 类提供的。它是一个功能性的控件，没有用户界面，主要用于在程序中按照时间间隔产生定时消息，然后执行消息代码。Timer 控件的常用属性和事件如表 4-24 所示。

表 4-24 Timer 控件的常用属性和事件

类别	名称	用途
属性	Name	指示代码中用来标识该对象的名称
	Interval	事件的频率（以毫秒为单位）
	Enabled	是否产生定时消息
事件	Tick	当经过指定的时间间隔时发生

4．状态栏控件

StatusStrip，即状态栏控件，是由 System.Windows.Forms.StatusStrip 类提供的，主要用于在应用程序中显示用户状态的简单信息，一般位于 Windows 窗体的底部。StatusStrip 控件的常用属性和事件如表 4-25 所示。

表 4-25 StatusStrip 控件的常用属性和事件

类别	名称	用途
属性	Name	指示代码中用来标识该对象的名称
	Text	显示的文本
	Items	显示项的集合
事件	ItemClicked	当单击项时发生

在状态栏中，可以使用文字或图标显示应用程序的状态，也可以用一系列图标组成动画来表示正在进行某个过程。在窗体中添加 StatusStrip 控件后，通过 Items 属性或单击右边的三角形按钮，将弹出一个下拉列表框，可以为状态栏添加 StatusLabel、ProgressBar、DropDownButton、SplitButton 等窗格控件，如图 4-21 所示。

图 4-21 在状态栏中添加窗格控件

这些控件的意义如表 4-26 所示。

表 4-26 状态栏中可以添加的控件

名称	说明
StatusLabel	表示 StatusStrip 控件中的一个面板
ProgessBar	显示进程的完成状态
DropDownButton	显示用户可以从中选择单个项关联的选项
SplitButton	表示作为标准按钮和下拉菜单的一个组合控件

【例 4-9】 设计一个窗体，在状态栏中显示当前时间。

【操作步骤】

（1）启动 VS，在 Windows 窗体应用程序 Controls 项目中新建一个窗体 Form8。

（2）双击 Form8.cs，切换到设计视图，从工具栏中添加 1 个 StatusStrip 控件和 1 个 Timer 控件到窗体设计区。

（3）在窗体设计区中右击窗体 Form8 和每一个控件，设置窗体和控件的相关属性。表 4-27 列出了窗体及控件属性。

表 4-27 窗体及控件属性设置

窗体和控件	属 性	属 性 值
Form8	Text	状态栏与定时器
statusStrip1: Items	Name	toolStripStatusLabel1
	Text	（空）
timer1	Name	timer1
	Enabled	True
	Interval	1000

（4）双击 Timer 控件，为其添加单击事件处理程序，程序代码如下：

```
private void timer1_Tick(object sender, EventArgs e)
{
    toolStripStatusLabel1.Text = DateTime.Now.ToString();
}
```

以上代码的功能是，每秒在状态栏中显示一次当前时间。

（5）在解决方案资源管理器 Controls 项目中双击 Program.cs 文件，将 Main()方法中的最后一行代码修改如下：

```
Application.Run(new Form8());
```

以上代码的功能是，Controls 项目运行时启动 Form8 窗体。

（6）编译并运行，运行结果如图 4-22 所示。

图 4-22 "状态栏与定时器"界面

5．进度条控件

ProgressBar，即进度条控件，是由 System.Windows.Forms.ProgressBar 类提供的，主要用于表示进度。ProgressBar 控件的常用属性和事件如表 4-28 所示。

表 4-28 ProgressBar 控件的常用属性和事件

类别	名称	用途
属性	Name	指示代码中用来标识该对象的名称
	Value	进度条的当前值
	Minimum	进度条的范围下限
	Maximum	进度条的范围上限
事件	Click	单击事件

【例 4-10】 设计一个进度条窗体，程序运行时，显示进度，当进度值达到 100 时，提示"下载完毕！"。

【操作步骤】

（1）启动 VS，在 Windows 窗体应用程序 Controls 项目中新建一个窗体 Form9。

（2）双击 Form9.cs，切换到设计视图，从工具栏中添加 1 个 Label、1 个 ProgressBar 控件和 1 个 Timer 控件到窗体设计区。

（3）在窗体设计区中右击窗体 Form9 和每一个控件，设置窗体和控件的相关属性。表 4-29 列出了窗体和控件的属性设置。

表 4-29 窗体和控件的属性设置

窗体和控件	属性	属性值
Form9	Text	进度条
label1	Text	下载进度
progressBar1	Name	progressBar1
	Maximum	100
	Minimum	0
timer1	Name	timer1
	Enabled	True
	Interval	100

（4）双击 Timer 控件，为其添加 Tick 事件处理程序，程序代码如下：

```
private void timer1_Tick(object sender, EventArgs e)
{
    progressBar1.Value++;
    if (progressBar1.Value == 100)
    {
        timer1.Stop();
        MessageBox.Show("下载完毕！");
    }
}
```

以上代码的功能是，每 0.1 秒进度条值加 1，当进度条的值达到 100 时，提示用户"下载完毕！"。

（5）在解决方案资源管理器 Controls 项目中双击 Program.cs 文件，将 Main()方法中的最后一行代码修改如下：

```
Application.Run(new Form9());
```

以上代码的功能是，Controls 项目运行时启动 Form9 窗体。

（6）编译并运行，运行结果如图 4-23 所示。

图 4-23 "进度条"界面

4.3 Windows 通用对话框

为了提高程序设计的效率，.NET 平台还封装了一些系统常用的对话框供开发者使用，这些对话框包括消息对话框、文件对话框以及普通对话框。使用.NET 提供的这些对话框控件，可以方便、快捷地实现一些常用功能。

4.3.1 消息对话框

C#中可以利用 MessageBox.Show()方法创建消息对话框，并利用 DialogResult 类型的变量接收返回值，以此来判断用户的操作行为或功能选项，进而执行相应的操作任务。消息对话框的语法格式如下：

MessageBox.Show(作用域,"对话框内容","对话框标题",按钮类型,图标类型)

【例 4-11】 设计一个窗体，当关闭窗体时创建消息对话框，单击"确定"按钮，则关闭当前窗体；否则，取消当前操作。

【操作步骤】
（1）启动 VS，在 Windows 窗体应用程序 Controls 项目中新建一个窗体 Form10。
（2）为窗体的 FormClosing 事件编写处理程序，程序代码如下：

```
private void Form10_FormClosing(object sender, FormClosingEventArgs e)
{
    DialogResult result = MessageBox.Show("确定要关闭吗？", "提示", MessageBoxButtons.OKCancel, MessageBoxIcon.Question);
    if (result == DialogResult.OK)
    {
        Form10 f = new Form10();
        f.Close();
    }
    else
    {
        e.Cancel = true;
    }
}
```

以上代码的功能是，当用户关闭窗体时，弹出提示框，如果用户单击"确定"按钮，则关闭 Form10 窗体；如果用户单击"取消"按钮，则关闭提示框。

（3）在解决方案资源管理器 Controls 项目中双击 Program.cs 文件，将 Main()方法中的最后一行代码修改如下：

```
Application.Run(new Form10());
```

以上代码的功能是，Controls 项目运行时启动 Form10 窗体。

（4）编译并运行，运行结果如图 4-24 所示。

图 4-24 "提示"对话框

4.3.2 文件对话框

文件对话框包括打开文件对话框 OpenFileDialog 和保存文件对话框 SaveFileDialog 等。

1．OpenFileDialog

OpenFileDialog 对应 System.Windows.Forms.OpenFileDialog 类，供用户打开或创建一个文件，在对话框中用户可以从整个磁盘目录及局域网上查找文件。OpenFileDialog 的常用属性和事件如表 4-30 所示。

表 4-30 OpenFileDialog 的常用属性和事件

类　别	名　称	用　途
属性	Name	指示代码中用来标识该对象的名称
	Title	对话框标题
	InitialDirectory	对话框显示的初始目录
	FileName	对话框的默认选择文件名
	Filter	文件类型（扩展名）过滤
事件	FileOK	对话框中打开按钮被单击
	HelpRequest	对话框中帮助按钮被单击

2．SaveFileDialog

SaveFileDialog 对应 System.Windows.Forms.SaveFileDialog 类，供用户保存文件时指定保存的位置或文件名。SaveFileDialog 的常用属性和事件如表 4-31 所示。

表 4-31 SaveFileDialog 的常用属性和事件

类别	名称	用途
属性	Name	指示代码中用来标识该对象的名称
	Title	对话框标题
	InitialDirectory	对话框显示的初始目录
	FileName	对话框的默认选择文件名
	Filter	文件类型（扩展名）过滤
事件	FileOK	对话框中打开按钮被单击
	HelpRequest	对话框中帮助按钮被单击

4.3.3 普通对话框

普通对话框包括颜色对话框 ColorDialog 和字体对话框 FontDialog 等。

1. ColorDialog

ColorDialog 对应 System.Windows.Forms.ColorDialog 类，供用户选择一种系统颜色或自定义颜色。ColorDialog 的常用属性如表 4-32 所示。

表 4-32 ColorDialog 的常用属性

类别	名称	用途
属性	Name	指示代码中用来标识该对象的名称
	Color	用户选择的颜色
	ShowHelp	是否显示帮助按钮
	AnyColor	是否可以选择任意颜色

2. FontDialog

FontDialog 对应 System.Windows.Forms.FontDialog 类，供用户选择字体。FontDialog 的常用属性如表 4-33 所示。

表 4-33 FontDialog 的常用属性

类别	名称	用途
属性	Name	指示代码中用来标识该对象的名称
	Color	颜色
	Font	被对话框选择的字体
	ShowEffects	是否显示下画线等选项

【例 4-12】简历编辑器。修改例 4-3 简历编辑器中的部分功能代码，使用 OpenFileDialog、SaveFileDialog、ColorDialog 和 FontDialog 对话框供用户进行简历编辑操作。

【操作步骤】

（1）启动 VS，打开 Controls 项目中的窗体 Form2。

（2）双击"加载"按钮，修改其单击事件处理程序，程序代码如下：

```
private void btnLoad_Click(object sender, EventArgs e)
{
```

```
OpenFileDialog f = new OpenFileDialog();
DialogResult re = f.ShowDialog();
if (re == DialogResult.OK)
{
    richTextBox1.LoadFile(f.FileName);
}
}
```

以上代码的功能是，当用户单击"加载"按钮时，启动"打开"对话框，通过对话框的 FileName 属性获得用户选择的文件名，当用户单击"打开"按钮时，RichTextBox 控件将加载用户选择的文件。"打开"对话框如图 4-25 所示。

图 4-25 "打开"对话框

（3）双击"保存"按钮，修改其单击事件处理程序，程序代码如下：

```
private void btnSave_Click(object sender, EventArgs e)
{
    SaveFileDialog f = new SaveFileDialog();
    DialogResult re = f.ShowDialog();
    if (re == DialogResult.OK)
    {
        richTextBox1.SaveFile(f.FileName);
    }
}
```

以上代码的功能是，当用户单击"保存"按钮时，启动保存文件对话框，通过对话框的 FileName 属性获得用户选择的文件名，RichTextBox 根据用户的选择保存文件。"另存为"对话框如图 4-26 所示。

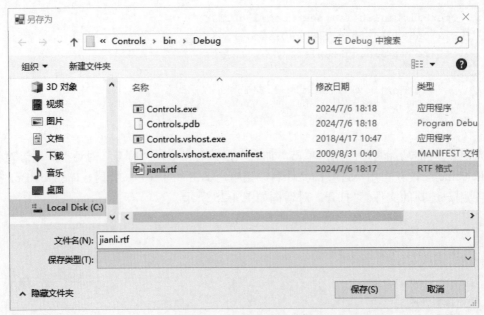

图 4-26 "另存为"对话框

（4）双击"颜色"按钮，修改其单击事件处理程序，程序代码如下：

```
private void btnColor_Click(object sender, EventArgs e)
{
    ColorDialog f = new ColorDialog();
    DialogResult re = f.ShowDialog();
    if (re == DialogResult.OK)
    {
        richTextBox1.SelectionBackColor = f.Color;
    }
}
```

以上代码的功能是，当用户单击"颜色"按钮时，启动"颜色"对话框，通过对话框的 Color 属性获得用户选择的颜色，RichTextBox 根据用户选择的颜色修改文字颜色。"颜色"对话框如图 4-27 所示。

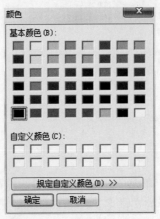

图 4-27 "颜色"对话框

(5) 双击"字体"按钮，修改其单击事件处理程序，程序代码如下：

```
private void btnFont_Click(object sender, EventArgs e)
{
    FontDialog f = new FontDialog();
    DialogResult re = f.ShowDialog();
    if (re == DialogResult.OK)
    {
        richTextBox1.SelectionFont = f.Font;
    }
}
```

以上代码的功能是，当用户单击"字体"按钮时，启动"字体"对话框，通过对话框的 Font 属性获得用户选择的字体，RichTextBox 根据用户的选择设置简历中的文字字体。"字体"对话框如图 4-28 所示。

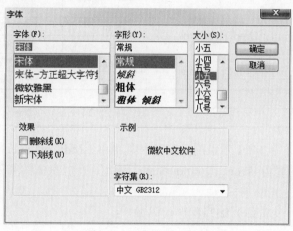

图 4-28 "字体"对话框

4.4 Windows 窗体设计

一个程序可能包含多个窗体，从对这些窗体的管理来看，C#程序设计通常可分为基于单文档界面的应用程序和基于多文档界面的应用程序两种。

4.4.1 基于单文档的窗体设计

基于单文档界面（Single Document Interface，SDI）应用程序，由单个文档、窗体或多个独立的窗体组成。单文档应用程序中每次只能处理一个当前激活的文档，如 Windows 操作系统中的画图工具和文本文档。本章前面的项目实例都属基于单文档的窗体设计。

4.4.2 基于多文档的窗体设计

基于多文档界面（Multiple Document Interface，MDI）应用程序中包含多个文档或窗体，所有文档都被组织到一个公共界面下。多文档的应用程序窗体通常分为父窗体和子窗体两种,包含其他窗体的窗体称为父窗体,只能被包含的窗体则称为子窗体。多文档应用程序可以同时打开多个子窗体,但只能有一个窗体处于激活状态,如 Microsoft Excel。

【例 4-13】 MDI 应用程序。

【设计步骤】

(1)启动 VS,在 Windows 窗体应用程序 Controls 项目中新建 3 个窗体 Form11、Form12 和 Form13。其中,Form11 为父窗体;Form12 和 Form13 为子窗体。

(2)双击 Form11.cs,切换到设计视图,为 Form11 添加 1 个 MenuStrip 控件到窗体设计区。

(3)设置 3 个窗体和控件的相关属性,表 4-34 列出了窗体和控件的属性设置。

表 4-34 窗体和控件的属性设置

窗体和控件	属 性	属 性 值
Form11	Text	MDI 应用
menuStrip1:Items	Name	formToolStripMenuItem
	Text	窗体
formToolStripMenuItem 1:DropDownItems	Name	form1ToolStripMenuItem
	Text	子窗体 1
formToolStripMenuItem 1:DropDownItems	Name	form2ToolStripMenuItem
	Text	子窗体 2
Form12	Text	子窗体 1
Form13	Text	子窗体 2

(4)设置 Form11 为 MDI 项目的父窗体。其中将 IsMdiContainer 属性(确定该窗体是否为 MDI 容器)设置为 True,WindowState 属性(确定窗体的初始可视状态)设置为 Maxmized。

(5)为 Form11(父窗体)编写程序代码如下:

```
namespace MDI
{
    public partial class Form11 : Form
    {
        public Form11()
        {
            InitializeComponent();
        }
        static bool formShow1 = false;        //标记子窗体1是否打开
        //添加属性FormShow1,用于关闭子窗体1时,修改标记
        static public bool FormShow1
        {
            get { return formShow1; }
            set { formShow1 = value; }
        }
        static bool formShow2 = false;        //标记子窗体2是否打开
        //添加属性FormShow2,用于关闭子窗体2时,修改标记
```

```
static public bool FormShow2
{
    get { return formShow2; }
    set { formShow2 = value; }
}
private void Form1ToolStripMenuItem_Click(object sender, EventArgs e)
{
    if (!formShow1)
    {
        Form12 f = new Form12();
        f.MdiParent = this;
        f.Show();
        formShow1 = true;
    }
}
private void Form2ToolStripMenuItem_Click(object sender, EventArgs e)
{
    if (!formShow2)
    {
        Form13 f = new Form13();
        f.MdiParent = this;
        f.Show();
        formShow2 = true;
    }
}
```

以上代码的功能是，定义两个属性 FormShow1 和 FormShow2，分别用来标记子窗体 1 和子窗体 2 是否已经被打开。当单击"子窗体 1"菜单时，打开且只能打开一次子窗体 1，当单击"子窗体 2"时，打开且只能打开一次子窗体 2。

（6）为 Form12（子窗体 1）的 FormClosing 事件编写处理程序，程序代码如下：

```
private void Form12_FormClosing(object sender, FormClosingEventArgs e)
{
    Form11.FormShow1 = false;
}
```

（7）为 Form13（子窗体 2）的 FormClosing 事件编写处理程序，程序代码如下：

```
private void Form13_FormClosing(object sender, FormClosingEventArgs e)
{
    Form11.FormShow2 = false;
}
```

（8）在解决方案资源管理器 Controls 项目中双击 Program.cs 文件，将 Main()方法中的最后一行代码修改如下：

```
Application.Run(new Form11());
```

以上代码的功能是，Controls 项目运行时启动 Form11 窗体。

（9）编译并运行，运行结果如图 4-29 所示。

图 4-29 "MDI 应用"对话框

习题4

第 5 章　输入与输出

学习目标：
- 理解文件与流的基本概念；
- 了解文件与目录的基本知识及其常用的类；
- 掌握文件流读写方法。

5.1　概　　述

文件是计算机管理数据的基本单位，同时也是应用程序保存和读取数据的一个重要场所。C#中文件处理技术称为 I/O 技术，即输入与输出技术，或称为流处理技术或文件流处理技术。

5.1.1　文件与流

文件和流是既有区别又有联系的两个概念。

文件是指在各种存储介质上永久保存数据的有序集合，并与一个具体的名称对应，它是进行文件读写操作的基本对象。从严格意义上讲，文件指的是放在磁盘上的静态信息，这种信息不是连续的，是随机的。

流是字节序列的抽象概念，流提供一种向后续存储器写入字节或从后续存储器读取字节的方法。流一般指的是连续的字节信息。例如，要对一个文件进行处理，这个文件就会变成连续的字节信息加载到内存中，也就是说文件在处理时就必须变成流。流强调的是动态的连续信息，是由文件转换成的。流和文件指的都是一件事物，但是状态不一样。

文件是存储在存储介质上的数据集，是静态的，具有名称和相应路径。当打开一个文件并对其进行读写时，该文件就成为流。

5.1.2　System.IO 命名空间

System.IO 命名空间包含允许读写文件和数据流的类型以及提供基本文件和目录支持的类型，因此在使用这些类时需要引入 System.IO 命名空间。System.IO 命名空间中的常用类如表 5-1 所示。

System.IO 命名空间中的常用类大致分为操作目录的类、操作文件的类、文件读写类等。其中，Directory 类和 DirectoryInfo 类属于操作目录的类；File 类和 FileInfo 类属于操作文件的类，StreamReader 类和 StreamWriter 类属于文本文件读写的类；BinaryReader 类和 BinaryWriter 类属于二进制文件读写的类。

表 5-1 System.IO 命名空间的类

类 名	功能和用途
Directory、DirectoryInfo	创建、删除并移动目录，通过属性获取特定目录的相关信息
File、FileInfo	创建、删除并移动文件，通过属性获取特定文件的相关信息
StreamReader、StreamWriter	读写文本数据信息
BinaryReader、BinaryWriter	读写二进制数据

5.2 目录操作

在程序开发中，有时需要对文件目录进行操作，如创建目录、删除目录等，为此 C# 提供了 Directory 类和 DirectoryInfo 类。

5.2.1 Directory 类

Directory 类是静态类，提供了许多静态方法用于对目录进行操作，如创建、删除和移动目录等。Directory 类的一些常用方法如表 5-2 所示。

表 5-2 Directory 类的常用方法

方 法	说 明
CreateDirectory()	创建指定路径的目录
Exists()	判断目录是否存在
GetDirectoryRoot()	获取指定目录的根目录
GetDirectories()	获取当前目录下的 Directory 对象数组
GetFiles()	获取当前目录下的 File 对象数组
Delete()	删除指定目录及其目录下的所有文件
Move()	将指定目录移动到新的位置

注意：Directory 的 Delete()方法是永久删除，不把目录送到回收站；使用 Move()方法移动目录时，要注意不能跨磁盘移动，如 C 盘的文件不能移到 D 盘下。

【例 5-1】 目录创建程序。

【操作步骤】

（1）启动 VS，新建一个 Windows 窗体应用程序 DirectoryApplication。

（2）双击 Form1.cs，切换到设计视图，从工具栏中拖曳 1 个 Label 控件、1 个 TextBox 控件和 1 个 Button 控件到窗体设计区，并调整控件大小进行布局。

（3）在窗体设计区中右击窗体 Form1 和每个控件，设置窗体和控件的相关属性。表 5-3 列出了窗体和控件的属性设置。

表 5-3 窗体和控件的属性设置

窗体和控件	属 性	属 性 值
Form1	Text	创建目录
label1	Name	请输入创建目录名称：
textBox1	Name	txtDirName
button1	Name	btnMake
	Text	创建

（4）双击"创建"按钮，为其添加单击事件处理程序，程序代码如下：

```
private void btnMake_Click(object sender, EventArgs e)
{
    string path = txtDirName.Text;
    if (Directory.Exists(path))
    {
        MessageBox.Show(path + "目录已经存在");
    }
    else
    {
        Directory.CreateDirectory(path);
        MessageBox.Show(path + "目录创建成功");
    }
}
```

以上代码的功能是，首先定义一个 string 类型的变量 path 保存用户输入的目录，然后调用 Directory 类的 Exists()方法查找该目录是否存在，Exists()方法的参数是用户输入的目录，即变量 path，如果存在，提示"目录已经存在"；如果不存在，则调用 Directory 类的 CreateDirectory()方法创建该目录，并提示"目录创建成功"。

（5）在解决方案资源管理器中右击 DirectoryApplication 项目，将其设为启动项目。

（6）编译并运行，运行结果如图 5-1 所示。

图 5-1　DirectoryApplication 项目运行结果

5.2.2　DirectoryInfo 类

DirectoryInfo 类的功能与 Directory 类相似，不同的是，DirectoryInfo 类是一个实例类，所有方法都是实例方法。也就是说，要想使用 DirectoryInfo 类提供的方法必须实例化一个属于 DirectoryInfo 类的对象。因此，如果需要对同一个目录进行多次重复操作时，应该考虑使用 DirectoryInfo 类的实例方法。DirectoryInfo 类不仅拥有与 Directory 类功能相似的方法，而且还具有一些特有的属性，如表 5-4 所示。

表 5-4　DirectoryInfo 类的属性

属　　性	说　　明
Name	获取当前 DirectoryInfo 对象的名称
Root	获取路径的根目录
Parent	获取指定子目录的父目录
FullName	获取目录或文件的完整目录
Exists	判断指定目录是否存在

【例 5-2】 目录浏览程序。

【操作步骤】

（1）启动 VS，新建一个 Windows 窗体应用程序 DirectoryInfoApplication。

（2）双击 Form1.cs，切换到设计视图，从工具栏中拖曳 2 个 GroupBox 控件、4 个 Label 控件、4 个 TextBox 控件、1 个 ListBox 控件和 1 个 Button 控件到窗体设计区，调整控件大小进行布局。

（3）在窗体设计区中右击窗体 Form1 和每一个控件，设置窗体和控件的相关属性。表 5-5 列出了窗体和控件的属性设置。

表 5-5 窗体和控件的属性设置

窗体和控件	属　性	属　性　值
Form1	Text	浏览目录
label1	Name	请输入目录名称：
textBox1	Name	txtDirName
label2	Text	目录名称：
textBox2	Name	txtName
label3	Text	根目录名称：
textBox3	Name	txtRoot
label4	Text	父目录名称：
textBox4	Name	txtParent
groupBox1	Text	详细信息
groupBox2	Text	目录列表
button1	Name	btnBrowse
	Text	浏览
listBox1	Name	listGetDirectories

（4）定义一个 dirList()方法，用于在列表框中循环输出指定目录中的所有子目录。

```
public void dirList(DirectoryInfo dir)
{
    listGetDirectories.Items.Clear();
    DirectoryInfo[] dirs = dir.GetDirectories();
    for (int i = 0; i < dirs.Length; i++)
    {
        listGetDirectories.Items.Add(dirs[i]);
    }
}
```

以上代码的功能是，封装一个 dirList()方法，该方法没有返回值，有一个 DirectoryInfo 类型的参数，方法体中首先清空列表框，并通过 DirectoryInfo 对象的 GetDirectories()方法获得目录中的所有子目录，所有子目录构成一个 DirectoryInfo 类型的数组，然后在列表框中循环输出该数组中的每个元素的值。

（5）双击"浏览"按钮，为其添加单击事件处理程序，程序代码如下：

```
private void btnBrowse_Click(object sender, EventArgs e)
```

```
{
    string path = txtDirName.Text;
    if (!string.IsNullOrEmpty(path))
    {
        DirectoryInfo dir = new DirectoryInfo(path);
        txtName.Text = dir.Name;
        txtRoot.Text = dir.Root.ToString();
        txtParent.Text = dir.Parent.ToString();
        //输出子目录
        dirList(dir);
    }
    else
    {
        MessageBox.Show("请输入目录名称！");
    }
}
```

以上代码的功能是，首先定义一个 string 类型的变量 path，用来存储用户输入的目录，然后通过调用 string 的 IsNullOrEmpty() 方法判断 path 值是否为空，即判断用户是否已输入目录名称。如果不为空，则创建该目录的 DirectoryInfo 对象，并在相应文本框中输出用户指定目录的名称、根目录名称、父目录名称，通过调用步骤（4）中封装的 dirList() 方法输出所有子目录。

（6）在解决方案资源管理器中右击 DirectoryInfoApplication 项目，将其设为启动项目。

（7）编译并运行，运行结果如图 5-2 所示。

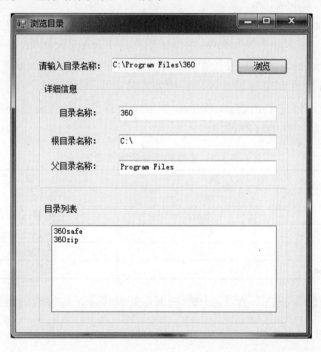

图 5-2　DirectoryInfoApplication 项目运行结果

5.3 文件操作

File 类和 FileInfo 类主要提供与文件有关的各种操作，包括创建、复制、移动、删除文件等。

5.3.1 File 类

File 类是一个静态类，提供了许多静态方法，用于处理文件。File 类的常用方法如表 5-6 所示。

表 5-6 File 类的常用方法

方 法	说 明
Create()	创建文件
Open()	打开指定路径上的文件，返回 FileStream 对象
Copy()	将文件复制到指定位置
Move()	将指定文件移动到新位置
Delete()	删除文件
Exists()	判断指定文件是否存在

注意：Directory 和 File 提供的方法都是共享方法，如果执行一次操作，使用共享方法的效率较高；但如果针对一个目录或文件多次操作，可以考虑使用 DirectoryInfo 和 FileInfo 提供的实例方法。

【例 5-3】 文件删除程序。

【操作步骤】

（1）启动 VS，新建一个 Windows 窗体应用程序 FileApplication。

（2）双击 Form1.cs，切换到设计视图，从工具栏中拖曳 1 个 Label 控件、1 个 TextBox 控件和 1 个 Button 控件到窗体设计区，并调整控件大小进行布局。

（3）在窗体设计区中右击窗体 Form1 和每一个控件，设置窗体和控件的相关属性。表 5-7 列出了窗体和控件的属性设置。

表 5-7 窗体和控件的属性设置

窗体和控件	属 性	属 性 值
Form1	Text	删除文件
label1	Name	请输入删除文件名称：
textBox1	Name	txtDirName
button1	Name	btnDelete
	Text	删除

（4）双击"删除"按钮，为其添加单击事件处理程序，程序代码如下：

```
private void btnDelete_Click(object sender, EventArgs e)
{
    string path = txtDirName.Text;
```

```
        if (File.Exists(path))
        {
            File.Delete(path);
            MessageBox.Show("文件删除完毕！");
        }
        else
        {
            MessageBox.Show ("文件不存在！");
        }
}
```

以上代码功能是，通过 File 类的 Exists()方法判断用户输入的文件是否存在，如果存在，通过调用 File 类的 Delete()方法删除该文件，并提示用户"文件删除完毕！"；如果不存在，提示"文件不存在！"。

（5）在解决方案资源管理器中右击 FileApplication 项目，将其设为启动项目。

（6）编译并运行，运行结果如图 5-3 所示。

图 5-3 FileApplication 项目运行结果

5.3.2 FileInfo 类

FileInfo 类与 File 类类似，它们都可以对磁盘上的文件进行操作。不同的是，FileInfo 类是实例类，所有的方法必须实例化对象后才能调用。FileInfo 类除了拥有与 File 类相似的方法外，同时也有它特有的属性，如表 5-8 所示。

表 5-8 FileInfo 类的属性

属　　性	说　　明
Directory	获取父目录的实例
DirectoryName	获取表示目录的完整路径的字符串
FullName	获取目录或文件的完整目录
Length	获取当前文件的大小

【例 5-4】 文件浏览程序。

【操作步骤】

（1）启动 VS，新建一个 Windows 窗体应用程序 FileInfoApplication。

（2）双击 Form1.cs，切换到设计视图，从工具栏中拖曳 1 个 GroupBox 控件、3 个 Label 控件、3 个 TextBox 控件和 1 个 Button 控件到窗体设计区，调整控件大小进行布局。

（3）在窗体设计区中右击窗体 Form1 和每一个控件，设置窗体和控件的相关属性。表 5-9 列出了窗体和控件的属性设置。

表 5-9　窗体和控件的属性设置

窗体和控件	属　　性	属　性　值
Form1	Text	浏览文件
label1	Name	请输入文件名称：
textBox1	Name	txtFileName
label2	Text	文件当前目录：
textBox2	Name	txtDir
label3	Text	文件大小：
textBox3	Name	txtSize
groupBox1	Text	详细信息
button1	Name	btnBrowse
	Text	浏览

（4）双击"浏览"按钮，为其添加单击事件处理程序，程序代码如下：

```
private void btnBrowse_Click(object sender, EventArgs e)
{
    FileInfo file = new FileInfo(txtFileName.Text);
    if (file.Exists)
    {
        txtDir.Text = file.Directory.ToString();
        txtSize.Text = file.Length.ToString();
    }
    else
    {
        file.Create();
        MessageBox.Show("文件已经创建成功！");
    }
}
```

以上代码的功能是，首先创建 FileInfo 类的对象 file，FileInfo 类的构造方法中有一个参数，即用户输入的文件名，通过调用对象 file 的 Exists 属性判断该文件是否存在。如果用户指定的文件存在，则输出文件的当前目录和文件大小；如果文件不存在，则通过调用对象的 Create()方法创建该文件，并提示"文件已经创建成功！"。

（5）在解决方案资源管理器中右击 FileInfoApplication 项目，将其设为启动项目。

（6）编译并运行，运行结果如图 5-4 所示。

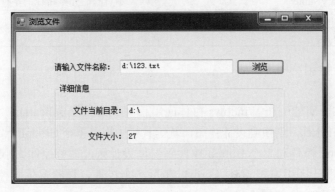

图 5-4　FileInfoApplication 项目运行结果

5.4 文件读写

数据流（也称为流）是一个用于传输数据的对象。数据的传输有两个方向，如果数据是从外部源传输到程序中，则称为读取流；如果数据是从程序传输到外部源，则称为写入流。

在 System.IO 命名空间中，对文件的读写操作是依靠流操作进行的。也就是说，.NET 中一个被打开的文件就是一个数据流，对已经打开的文件进行读写操作就是在对内存中的一个数据流进行读写操作。

5.4.1 读写文本文件

文本文件是一种纯文本数据构成的文件。实际上，文本文件只保存了字符的编码。.NET Framework 支持多种编码，包括 ASCII、UTF7、UTF8、Unicode、UTF32 等。在.NET Framework 中，读写文本文件可以使用流读取器 StreamReader 和流写入器 StreamWriter。

1. StreamReader 类

StreamReader 类用于从文件中读取数据，该类是一个通用类，可用于任何流。StreamReader 类以一种特定的编码输入字符，默认的编码为 UTF8，UTF8 可以正确处理 Unicode 字符并在操作系统的本地化版本上提供一致的结果。StreamReader 类的常用方法如表 5-10 所示。

表 5-10 StreamReader 类的常用方法

方法	说明
Close()	关闭 StreamReader 对象和基础流
Dispose()	释放所有 StreamReader 对象资源
Peek()	返回下一个可用的字符
Read()	读取输入流中的下一个字符或下一组字符
ReadLine()	从数据流中读取一行数据，并作为字符串返回
ReadToEnd()	从流的当前位置到末尾读取流

2. StreamWriter 类

StreamWriter 类用于将字符和字符串写入文件，它实际上也是先转换成 FileStream 对象，然后再向文件中写入数据的，所以在创建对象时可以通过 FileStream 对象来创建 StreamWriter 对象，同时也可以直接创建 StreamWriter 对象。StreamWriter 默认使用 UTF8 编码。StreamWriter 类的常用方法如表 5-11 所示。

表 5-11 StreamWriter 类的常用方法

方法	说明
Close()	关闭 StreamWriter 对象和基础流
Dispose()	释放所有 StreamWriter 对象资源
Flush()	清理当前编写器的所有缓冲区，并使所有缓冲数据写入基础流
Write()	写入流
WriteLine()	写入指定的某些数据，后跟行结束符

【例 5-5】 简单日志程序。

【操作步骤】

(1) 启动 VS，新建一个 Windows 窗体应用程序 StreamApplication。

(2) 双击 Form1.cs，切换到设计视图，从工具栏中拖曳 2 个 Label、2 个 TextBox 控件和 2 个 Button 控件到窗体设计区，调整控件大小进行布局。

(3) 在窗体设计区中右击窗体 Form1 和每个控件，设置窗体和控件的相关属性。表 5-12 列出了窗体和控件的属性设置。

表 5-12 窗体和控件的属性设置

窗体和控件	属　性	属　性　值
Form1	Text	简单日志
label1	Name	请输入日志内容：
textBox1	Name	txtSource
	MultiLine	true
label2	Text	已有的日志内容：
textBox2	Name	txtShow
	MultiLine	true
	ReadOnly	true
button1	Name	btnSave
	Text	保存
button2	Name	btnShow
	Text	显示

(4) 双击"保存"按钮，为其添加单击事件处理程序，程序代码如下：

```
private void btnSave_Click(object sender, EventArgs e)
{
    StreamWriter writer = new StreamWriter(@"D:\示例代码\chapter05\日志.txt",true);
    writer.WriteLine(DateTime.Now.ToString());
    writer.WriteLine(txtSource.Text);
    writer.Close();
    MessageBox.Show("日志保存成功！");
}
```

以上代码的功能是，在保存日志时，首先利用 StreamWriter 类的构造方法创建流写入器对象，构造方法的第 1 个参数表示文件名的路径；第 2 个参数表示是否添加新内容，如果设置为 false，将覆盖原有内容。然后调用 WriteLine()方法把日志内容写入文件。

(5) 双击"显示"按钮，为其添加单击事件处理程序，程序代码如下：

```
private void btnShow_Click(object sender, EventArgs e)
{
    StreamReader reader = new StreamReader(@"D:\示例代码\chapter05\日志.txt");
    txtShow.Text = reader.ReadToEnd();
    reader.Close();
}
```

以上代码的功能是，在读取日志内容时，首先利用 StreamReader 类的构造方法创建读取流的读取器对象，同时打开磁盘文件，接着调用 ReadToEnd()方法，把文件内容全部读出，返回的字符串通过文本框输出。

（6）在解决方案资源管理器中右击 StreamApplication 项目，将其设为启动项目。

（7）编译并运行，运行结果如图 5-5 所示。

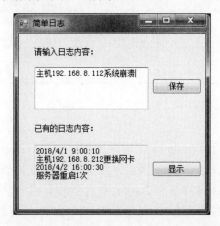

图 5-5　StreamApplication 项目运行结果

5.4.2　读写二进制文件

在.NET Framework 中，读写二进制文件可以使用流读取器 BinaryReader 和流写入器 BinaryWriter。二进制文件是以二进制代码形式存储的文件，数据存储为字节序列。二进制文件可以包含图像、声音、文本或编译之后的程序代码。

C#的 FileStream 类提供了最原始的字节级上的文件读写功能，但编程中经常会对字符串操作，于是 StreamWriter 类和 StreamReader 类增强了 FileStream 类，它可以在字符串级别上操作文件。但有时还是需要在字节级上操作文件，却又不是一字节一字节地操作，通常是 2、4 或 8 字节操作，这便有了 BinaryWriter 和 BinaryReader 类，它们可以将一个字符或数字按指定个数字节写入，也可以一次读取指定字节转为字符或数字。

1. BinaryReader 类

BinaryReader 类用特定的编码将基元数据类型读作二进制值，其常用方法如表 5-13 所示。

表 5-13　BinaryReader 类的常用方法

方法	说明
Close()	关闭 BinaryReader 对象和基础流
Dispose()	释放 BinaryReader 类当前实例所使用的所有资源
PeekChar()	返回下一个可用的字符，并且不提升字节或字符的位置
Read()	从基础流中读取字符，并根据所使用的 Encoding 和从流中读取的特定字符，提升流的当前位置
ReadByte()	从当前流中读取下一字节，并使流的当前位置提升一字节
ReadString()	从当前流中读取一个字符串。字符串有长度前缀，一次 7 位地被编码为整数

2. BinaryWriter 类

BinaryWriter 类以二进制形式将基元数据类型写入流,并支持用特定的编码写入字符串,其常用方法如表 5-14 所示。

表 5-14 BinaryWriter 类的常用方法

方 法	说 明
Close()	关闭当前 BinaryWriter 和基础流
Dispose()	释放 BinaryWriter 类当前实例所使用的所有资源
ToString()	返回表示当前对象的字符串
Write()	将值写入当前流

【例 5-6】 简单学生管理程序。

【操作步骤】

(1) 启动 VS,新建一个 Windows 窗体应用程序 BinaryApplication。

(2) 双击 Form1.cs,切换到设计视图,从工具栏中拖曳 3 个 Label、2 个 TextBox 控件、2 个 RadioButton、1 个 ListBox 和 2 个 Button 控件到窗体设计区,调整控件大小进行布局。

(3) 在窗体设计区中右击窗体 Form1 和每一个控件,设置窗体和控件的相关属性。表 5-15 列出了窗体和控件的属性设置。

表 5-15 窗体和控件的属性设置

窗体和控件	属 性	属 性 值
Form1	Text	学生信息管理
label1	Name	学号:
textBox1	Name	txtNum
label2	Text	姓名:
textBox2	Name	txtName
label3	Text	性别:
radioButton1	Name	rdbMale
	Text	男
radioButton2	Name	rdbFemale
	Text	女
listBox1	Name	listShow
button1	Name	btnSave
	Text	保存
button2	Name	btnShow
	Text	显示

(4) 双击"保存"按钮,为其添加单击事件处理程序,程序代码如下:

```
private void btnSave_Click(object sender, EventArgs e)
{
    FileStream fs = new FileStream(@"D:\示例代码\chapter05\student.dat",
    FileMode.Append ,FileAccess .Write);
    //通过文件流写文件
    BinaryWriter writer = new BinaryWriter(fs);
```

```
        //写入一个整数
        writer.Write(Int32.Parse(txtNum.Text));
        //写入一个字符串
        writer.Write(txtName.Text);
        bool isMale;
        if (rdbMale.Checked)
        {
            isMale = true;
        }
        else
        {
            isMale = false;
        }
        //写入一个bool值
        writer.Write(isMale);
        fs.Close();
        writer.Close();
    }
```

以上代码的功能是,在保存数据时,首先利用 FileStream 类的构造方法创建一个文件流对象,该构造方法有 3 个参数,第 1 个参数表示要操作的文件名;第 2 个参数是文件模式,FileMode.Append 表示打开现有文件并查找到文件尾,如果文件不存在,则创建该文件;第 3 个参数是文件操作模式,FileAccess.Write 表示写文件。然后通过文件流对象创建 BinaryWriter 写入器对象,并连续调用写入器对象的 Write()方法把数据写入文件流。

(5)双击"显示"按钮,为其添加单击事件处理程序,程序代码如下:

```
    private void btnShow_Click(object sender, EventArgs e)
    {
        listShow.Items.Clear();
        listShow.Items.Add("学号\t姓名\t性别");
        FileStream fs = new FileStream(@"D:\示例代码\chapter05\student.dat",
        FileMode.Open,FileAccess.Read);
        //通过文件流读文件
        BinaryReader reader = new BinaryReader(fs);
        fs.Position = 0;
        while (fs.Position !=fs.Length)
        {
            //读出一个整数
            int num = reader.ReadInt32();
            //读出一个字符串
            string name = reader.ReadString();
            string sex = "";
            //读出一个bool值
            if (reader.ReadBoolean())
            {
```

```
            sex = "男";
        }
        else
        {
            sex = "女";
        }
        string result = string.Format("{0}\t{1}\t{2}", num, name, sex);
        listShow.Items.Add(result);
    }
    reader.Close();
    fs.Close();
}
```

以上代码的功能是，在显示数据时，首先创建文件流对象，并指定操作方式为打开（FileMode.Open）和读取文件（FileAccess.Read），然后通过文件流对象创建 BinaryReader 读取器对象，并使用读取器对象从头至尾循环读取文件流，最终把读出来的数据添加到列表框中输出。

（6）在解决方案资源管理器中右击 BinaryApplication 项目，将其设为启动项目。

（7）编译并运行，运行结果如图 5-6 所示。

图 5-6 BinaryApplication 项目运行结果

习题5

第 6 章　数据访问技术

学习目标：
- 了解 ADO.NET 的基本知识；
- 掌握断开式和非断开式操作数据库方法；
- 掌握使用 Connection 对象连接数据库的方法；
- 掌握使用 Command 对象执行数据库命令的方法；
- 掌握使用 DataAdapter 对象执行数据库命令的方法。

6.1　数据库基础

数据库管理已经成为现代管理信息系统强有力的工具，最流行的数据库就是关系数据库，而 SQL 则是操作数据库的通用标准语言。

6.1.1　数据库的基本概念

数据库（Database）是计算机中存储数据的仓库，是一个由一批数据构成的有序集合，这个集合通常保存为一个或多个彼此相关的文件，这些数据分类别地存放在一些结构化的数据表中。

关系数据库采用现代数学理论和方法对数据进行处理，提供了结构化查询语言 SQL，操作和应用十分方便。关系数据库把数据组成一张或多张二维的表格，多张彼此相关联的表格群即组成数据库。关系数据库使用字段、记录、数据表、数据库、主键等术语，其意义描述如下：

（1）字段（Field）：二维表中的每列称为一个字段，用于描述关系的属性特征。用字段名来区分不同的字段，每个字段的字段名、数据类型、长度等是在创建表时规定的。例如，表 6-1 中的 num、name、gender 等都是字段。

（2）记录（Record）：二维表中由各字段取值构成的每一行数据称为一条记录。例如，表 6-1 中的"1，李雷，1"就是一条记录。

（3）数据表（Table）：由字段和记录所组成的一个没有重复行和列的二维表格称为一个关系数据表。例如，表 6-1 studentinfo 就是数据表。

（4）数据库（Database）：多个相关联的数据表的集合构成一个数据库。例如，表 6-1 的学生信息表和表 6-2 学生成绩表描述的是有关学生和成绩的关系，因此可将它们组成一个数据库（studentmanage）。

（5）主键（Key）：能唯一地标识不同记录的单个或多个字段的组合称为主键。例如，表 6-1 中的学号 num。

表 6-1 学生信息表 studentinfo

num	name	gender
1	李雷	0
2	刘小红	1
3	张强	0

表 6-2 学生成绩表 studentscore

num	name	score
1	李雷	87
2	刘小红	97
3	张强	76

注意：可以使用 MySQL 创建数据库 studentmanage，并添加学生信息表（studentinfo）和学生成绩表（studentscore），用于本章实例操作。

6.1.2 数据库访问过程

一个典型的数据库访问过程如图 6-1 所示。

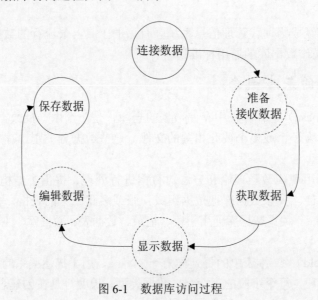

图 6-1 数据库访问过程

1．连接数据

连接数据为必选步骤。为了将数据引入应用程序，需要建立双向通信机制。这种双向通信机制由一个连接对象处理，通过连接对象配置连接到数据源时所需要的信息（连接字符串）。

2．准备接收数据

准备接收数据为可选步骤。只有当应用程序采用断开式访问数据库时，在处理数据期间才需要在应用程序中临时存储数据，因此，在获取数据之前，需要创建一个数据集，用以接收数据。

3．获取数据

获取数据为必选步骤。通过对数据库执行查询或存储过程将数据引入应用程序。

4．显示数据

显示数据为可选步骤。在将数据引入应用程序后，可以将它显示在窗体上供用户查看或修改。

5．编辑数据

编辑数据为可选步骤。获取数据后，用户可能会对数据进行添加、修改、删除等操作。

6．保存数据

保存数据为必选步骤。将操作后的数据返回数据库。

6.2　ADO.NET

微软公司在.NET编程环境中优先使用的数据访问接口ADO.NET提供了平台互用性和可伸缩的数据访问，为与数据库相关的应用程序开发提供了极大的便利。

6.2.1　ADO.NET 概述

ADO.NET 是.NET 框架的组件，是在基于.NET 平台的应用程序中用于访问数据源的技术，也是.NET Framework 提供的数据访问类库。应用程序可以使用 ADO.NET 连接到这些数据源，并检索和更新所包含的数据。ADO.NET 功能示意图如图 6-2 所示。

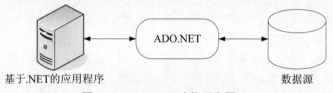

图 6-2　ADO.NET 功能示意图

ADO.NET 体系结构如图 6-3 所示，其中包含.NET Framework 数据提供程序和 DataSet 两大核心组件。

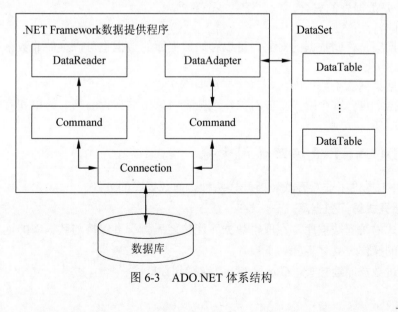

图 6-3　ADO.NET 体系结构

1. .NET Framework 数据提供程序

.NET Framework 数据提供程序包括以下几种。

（1）SQL Server .NET Framework 数据提供程序，提供对 Microsoft SQL Server 7.0 版本或更高版本的数据访问。

（2）OLE DB .NET Framework 数据提供程序，适用于使用 OLE DB 公开的数据源。

（3）ODBC .NET Framework 数据提供程序，适用于使用 ODBC 公开的数据源。

（4）Oracle .NET Framework 数据提供程序，适用于 Oracle 数据源。

在 ADO.NET 中，提供了 4 种连接数据源的接口，分别是 SQLClient、OracleClient、OLE DB 和 ODBC。其中，SQLClient 是 Microsoft SQL Server 数据库专用连接接口；OracleClient 是 Oracle 数据库专用的连接接口；ODBC 和 OLE DB 可用于其他数据源的连接。在应用程序中使用任何一种连接接口时，必须在后台代码中引用对应的名称空间，类的名称也随之发生变化，如表 6-3 所示。

表 6-3 数据连接方式名称空间与对应的类名称

名称空间	对应的类名称
System.Data.SqlClient	SqlConnection、SqlCommand、SqlDataReader、SqlDataAdaper
System.Data.OleDb	OleDbConnection、OleDbCommand、OleDbDataReader、OleDbDataAdaper
System.Data.Odbc	OdbcConnection、OdbcCommand、OdbcDataReader、OdbcDataAdaper
System.Data.OracleClient	OracleConnection、OracleCommand、OracleDataReader、OracleDataAdaper

.NET Framework 数据提供程序包含以下 4 个核心类。

（1）Connection：建立与数据源的连接。

（2）Command：对数据源执行操作命令，用于修改数据、查询数据、运行存储过程等。

（3）DataReader：从数据源获取只读只进的数据流。

（4）DataAdaper：用数据源数据填充 DataSet，并可以处理数据更新。

2. DataSet

DataSet 是 ADO.NET 的断开式结构的核心组件。设计 DataSet 的目的是实现独立于任何数据源的数据访问，可以把它看成内存中的数据库，是专门用来处理数据库中读出的数据。DataSet 是数据表的集合，包含任意多个数据表。

数据集的特点：

（1）数据集是存储数据的对象，可以存储数据库的数据也可以存储非数据库的数据，如 XML 中的数据。

（2）数据集独立于数据库，不直接与数据库交互。

（3）数据集包含零个或多个表对象，这些表对象由数据行与列、约束和有关表中数据关系的信息组成。

6.2.2 ADO.NET 数据库访问步骤

ADO.NET 提供了非断开式和断开式两种访问数据库的方式。

1. 非断开式访问数据库

非断开式的访问数据库，在取得数据库连接之后将一直保持与数据库的连接，直到执行关闭连接的操作。具体步骤如下：

（1）通过数据库连接类（Connection）连接到数据库。

（2）通过数据库命令类（Command）在数据库上执行 SQL 语句，实现对数据库的插入（Insert）、删除（Delete）、更新（Update）、查询（Select）等操作。

（3）如果是查询操作，可以通过数据库读取器类（DataReader）进行数据记录的向前只读操作。

（4）数据库操作完成后，再通过连接类（Connection）关闭当前的数据连接。

在非断开式访问数据库过程中，由于数据库的客户连接数量有限，因此，应尽量缩短与数据库操作的时间，数据库操作一旦完成，应及时关闭对应的数据库连接。

2．断开式访问数据库

断开式访问数据库，当数据适配器从数据源得到数据，或是从数据集更新数据到数据源后，其与数据源的连接就马上断开。具体步骤如下：

（1）通过数据库连接类（Connection）连接到数据库。

（2）创建基于当前数据库连接的数据适配器（DataAdaper），数据适配器通过数据库命令类（Command）从数据库获取数据到本地数据集（DataSet）中。

（3）对本地数据集中的数据进行插入、删除、更新、查询等操作。

（4）通过适配器将本地数据更新到数据库。

断开式访问数据库方式，具有执行效率高、数据库连接占用时间短等优点，但也存在数据更新不及时的缺点。

6.3　ADO.NET 数据库访问操作

本书中案例选用 MySQL 数据库，ADO.NET 访问 MySQL 数据库时，需要进行如下操作。

（1）在解决方案资源管理器中，右击该项目，选择"引用"→"管理 NuGet 程序包"选项，打开 NuGet 包管理器，在搜索框中输入"mysql.data"，如图 6-4 所示。

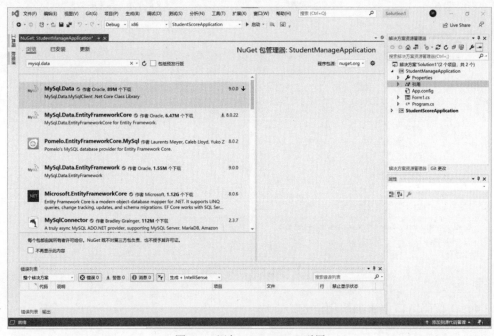

图 6-4　添加 MySql.Data 引用

(2)单击↓进入安装界面,如图 6-5 所示。

图 6-5 MySql.Data 安装

(3)安装完成后可在程序包管理器"已安装"中查看,如图 6-6 所示。

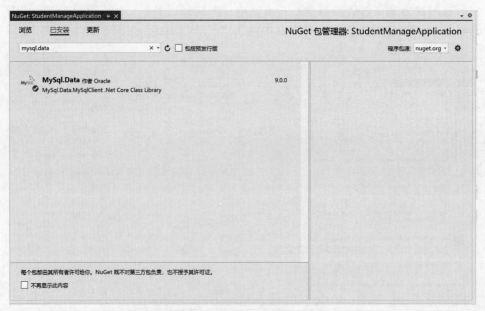

图 6-6 MySql.Data 安装完成

(4)在代码窗口中添加引用语句:

```
using MySql.Data.MySqlClient;
```

6.3.1 使用 Connection 对象连接数据库

创建数据库连接，使用 Connection 对象（当使用 MySQL 数据库时，为 MySqlConnection 对象），Connection 的构造方法和方法描述如下。

1．构造方法（使用一个参数的）

Connection（ConnectionString）：ConnectionString 是连接字符串，连接字符串的参数主要包括服务器、数据库用户名、数据库密码、数据库名。

构造连接字符串有以下两种方法（假定连接字符串名称为 conStr）。

（1）直接构造，格式如下：

```
string conStr = "server = 服务器名称;user id = 登录数据库的用户名;password = 登录数据库的密码;database = 数据库名称";
```

（2）在配置文件中存储连接字符串，格式如下：

```
<connectionStrings>
    <add name = "连接字符串名称" connectionString = "server = 服务器名称;user id = 登录数据库的用户名;password = 登录数据库的密码;database = 数据库名称" />
</connectionStrings>
```

在页面获取配置文件中的连接字符串，格式如下：

```
string conStr = ConfigurationManager.ConnectionStrings["连接字符串名称"].ConnectionString;
```

2．方法

（1）Open()：打开数据库。
（2）Close()：关闭数据库。

【例 6-1】 连接数据库。

【操作步骤】

（1）启动 VS，新建一个 Windows 窗体应用程序 StudentManageApplication。

（2）双击 Form1.cs，切换到设计视图，从工具栏中拖曳一个 Button 控件到窗体设计区，并调整控件大小进行布局。

（3）在窗体设计区中右击窗体 Form1 和每一个控件，设置窗体和控件的相关属性。表 6-4 列出了窗体和控件的属性设置。

表 6-4 窗体和控件的属性设置

窗体和控件	属　性	属　性　值
Form1	Text	学生信息管理
button1	Name	btnConnect
	Text	连接数据库

（4）右击 StudentManageApplication 项目，添加 MySql.Data.dll 引用，并在代码窗口中添加引用语句，程序代码如下：

```
using MySql.Data.MySqlClient;
```

（5）双击"连接数据库"按钮，为其添加单击事件处理程序，程序代码如下：

```
private void btnConnect_Click(object sender, EventArgs e)
{
    //创建连接字符串
    string conStr = "server=localhost;user id=root;password=123456;database=studentmanage";
    MySqlConnection con = new MySqlConnection(conStr);
    con.Open();
    MessageBox.Show("数据库连接成功！");
    con.Close();
}
```

以上代码的功能是，首先定义连接字符串 conStr，其中 4 个参数分别是服务器名称、登录数据库的用户名、登录数据库的密码以及数据库名称；然后创建一个 MySqlConnection 对象，其中连接字符串 conStr 作为构造方法的参数，通过 MySqlConnection 对象调用 Open() 方法来打开数据库，并提示用户"数据库连接成功！"；最后关闭连接。

（6）在解决方案资源管理器中右击 StudentManageApplication 项目，将其设为启动项目。

（7）编译并运行，运行结果如图 6-7 所示。

图 6-7 StudentManageApplication 项目运行结果

6.3.2 使用 Command 对象执行数据库命令

Command 对象（当使用 MySQL 数据库时，为 MySqlCommand 对象）是用来执行数据库操作命令的，如数据库中数据表记录的查询、增加、删除和修改等。一个数据库操作命令可以用 SQL 语句来表达，包括 select 语句、insert 语句、delete 语句和 update 语句等。Command 对象可以传递参数并返回值，同时 Command 也可以调用数据库中的存储过程。Command 的构造方法和方法描述如下。

（1）构造方法。

Command(CommandText,Connection)：CommandText 是要执行的 SQL 命令；Connection 为使用的数据连接对象。

（2）方法。

① ExecuteNonQuery()：返回的是被影响的行数（整数），一般在修改时使用，不会造成内存泄漏。

② ExecuteReader()：返回的是 MySqlDataReader 对象。

1．使用 Command 对象查询数据

使用 Command 对象查询数据库中数据时，先建立数据库连接；然后创建 Command 对象，并设置它的 CommandText 和 Connection 属性；接着使用 Command 对象的 ExecuteReader() 方法，把返回结果放在 DataReader 对象中；最后通过循环，处理数据库查询结果。

【例 6-2】 查询学生信息。

【操作步骤】

（1）启动 VS，打开 Windows 窗体应用程序 StudentManageApplication。

（2）双击 Form1.cs，切换到设计视图，修改 Form1 窗体，从工具栏中拖曳 3 个 Label 控件、2 个 TextBox 控件、2 个 RadioButton 控件和 2 个 Button 控件到窗体设计区，并调整控件大小进行布局。

（3）在窗体设计区中右击窗体 Form1 和每一个控件，设置窗体和控件的相关属性。表 6-5 列出了新增控件属性。

表 6-5 控件属性设置

控 件	属 性	属 性 值
label1	Text	学号：
textBox1	Name	txtNum
	ReadOnly	true
label2	Text	姓名：
textBox2	Name	txtName
label3	Text	性别：
radioButton1	Name	rdbMale
	Text	男
radioButton2	Name	rdbFemale
	Text	女
button2	Name	btnPrevious
	Text	上一个
button3	Name	btnNext
	Text	下一个

（4）由于本项目中需要多次使用连接字符串，因此可以使用 6.3.1 节中介绍的第二种方法存储连接字符串，减少代码量。右击项目，添加应用程序配置文件 App.config，并设置连接字符串的值，程序代码如下：

```
<connectionStrings>
    <add name = "conStr" connectionString = "server = localhost;user id = root;password = 123456;database = studentmanage;Charset = utf8;" />
</connectionStrings>
```

（5）为项目添加引用，用于在配置文件中读取连接字符串，程序代码如下：

using System.Configuration;

（6）在 Form1 类中添加成员字段 current 和 conStr，分别用于保存当前显示的学生学号和数据库连接字符串，程序代码如下：

```
private int current = 1;
string conStr = ConfigurationManager.ConnectionStrings["conStr"].ConnectionString;
```

（7）在 Form1 类中封装一个 ShowCurrentStudent()方法，用于查询和显示当前学生，程序代码如下：

```
private void ShowCurrentStudent()
{
    string sql = "select * from studentinfo where num="+current;
    MySqlConnection con = new MySqlConnection(conStr);
    con.Open();
    //创建Command对象
    MySqlCommand cmd = new MySqlCommand(sql, con);
    MySqlDataReader reader = cmd.ExecuteReader();
    //读取数据行
    if (reader.Read())
    {
      //显示学生学号
       txtNum.Text = reader.GetInt32(0).ToString();
      //显示学生姓名
       txtName.Text = reader.GetString(1);
      //显示学生性别
       string gender = reader.GetString(2);
       if (gender=="0")
       {
           rdbMale.Checked = true;
       }
       else
       {
           rdbFemale.Checked = true;
       }
    }
    reader.Close();
    con.Close();
}
```

以上代码的功能是，封装一个 ShowCurrentStudent()方法，用于查询并显示学生信息，该方法没有返回值也没有参数。在方法体中首先定义一个 SQL 语句，用于在数据库表 studentinfo 中查询学号 num 值为 current 变量值的数据，即当前学生信息；然后通过调用 Command 对象的 ExecuteReader()方法读取数据，并将第 1 个字段的值显示到学号文本框，第 2 个字段的值显示到学生姓名文本框，第 3 个字段的值如果是 0，则表示性别是"男"，否则性别为"女"。

(8) 双击 Form1 窗体,为其添加窗体加载事件处理程序,程序代码如下:

```
private void Form1_Load(object sender, EventArgs e)
{
    ShowCurrentStudent();
}
```

以上代码的功能是,当窗体加载时,将调用已封装的 ShowCurrentStudent()方法,即在窗体中显示当前学生信息,即学号为 1 的学生信息。

(9) 双击"上一个"按钮,为其添加单击事件处理程序,程序代码如下:

```
private void btnPrevious_Click(object sender, EventArgs e)
{
    current--;
    ShowCurrentStudent();
}
```

以上代码的功能是,当前学生的学号值减 1,调用已封装的 ShowCurrentStudent()方法查询并显示学生信息。

(10) 双击"下一个"按钮,为其添加单击事件处理程序,程序代码如下:

```
private void btnNext_Click(object sender, EventArgs e)
{
    current++;
    ShowCurrentStudent();
}
```

以上代码的功能是,当前学生的学号值加 1,调用已封装的 ShowCurrentStudent()方法查询并显示学生信息。

(11) 编译并运行,运行结果如图 6-8 所示。

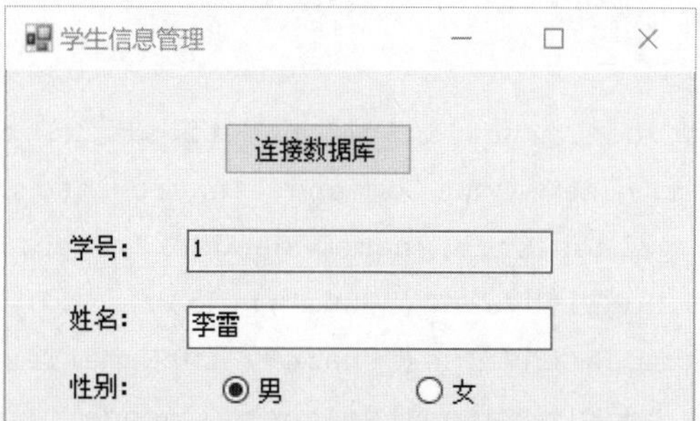

图 6-8 StudentManageApplication 项目运行结果

2. 使用 Command 对象增加、修改和删除数据

使用 Command 对象增加、修改和删除数据库数据时，首先建立数据库连接；然后创建 Command 对象，设置它的 CommandText 和 Connection 属性，并使用 Command 对象的 Parameters 属性设置输入参数；最后使用 Command 对象的 ExecuteNonQuery()方法执行数据增加、修改和删除命令。

【例 6-3】增加、修改和删除学生信息。

【操作步骤】

（1）启动 VS，打开 Windows 窗体应用程序 StudentManageApplication。

（2）双击 Form1.cs，切换到设计视图，修改 Form1 窗体，从工具栏中拖曳 3 个 Button 控件到窗体设计区，并调整控件大小进行布局。

（3）在窗体设计区中右击窗体 Form1 和每一个控件，设置窗体和控件的相关属性。表 6-6 列出了新增控件属性。

表 6-6 控件属性设置

控件	属性	属性值
button4	Name	btnInsert
	Text	增加
button5	Name	btnUpdate
	Text	修改
button6	Name	btnDelete
	Text	删除

（4）双击"增加"按钮，为其添加单击事件处理程序，程序代码如下：

```csharp
private void btnInsert_Click(object sender, EventArgs e)
{
    string gender = "0";
    if (rdbFemale.Checked)
    {
        gender = "1";
    }
    MySqlConnection con = new MySqlConnection(conStr);
    MySqlCommand cmd = new MySqlCommand("insert into studentinfo (num,name,
        gender) values(@num,@name,@gender)", con);
    cmd.Parameters.AddWithValue("@num",null);
    cmd.Parameters.AddWithValue("@name", txtName.Text);
    cmd.Parameters.AddWithValue("@gender", gender);
    con.Open();
    cmd.ExecuteNonQuery();
    con.Close();
    MessageBox.Show("成功增加一条数据!");
}
```

以上代码的功能是，使用 MySqlCommand 对象实现数据的添加，其中 Command 对象中 SQL 语句使用的参数用"@参数名"进行标识。例如，语句"MySqlCommand cmd = new MySqlCommand("insert into studentinfo(num,name,gender) values(@num,@name,@gender)", con);"中的@num、@name 和@gender 都是参数，分别代表学生的 3 项信息。定义参数后，需要将参数添加到 Command 对象的参数集合中并给参数赋值。例如，"cmd.Parameters.AddWithValue("@name", txtName .Text);"是将姓名文本框的值赋给参数@name。

（5）双击"修改"按钮，为其添加单击事件处理程序，程序代码如下：

```
private void btnUpdate_Click(object sender, EventArgs e)
{
    string gender = "0";
    if (rdbFemale.Checked)
    {
        gender = "1";
    }
    MySqlConnection con = new MySqlConnection(conStr);
    MySqlCommand cmd = new MySqlCommand("update studentinfo set name = @name,gender = @gender where num = @num", con);
    cmd.Parameters.AddWithValue("@name", txtName.Text);
    cmd.Parameters.AddWithValue("@gender", gender);
    cmd.Parameters.AddWithValue("@num", Convert.ToInt32(txtNum.Text));
    con.Open();
    cmd.ExecuteNonQuery();
    con.Close();
    MessageBox.Show("成功修改一条记录!");
}
```

（6）双击"删除"按钮，为其添加单击事件处理程序，程序代码如下：

```
private void btnDelete_Click(object sender, EventArgs e)
{
    MySqlConnection con = new MySqlConnection(conStr);
    MySqlCommand cmd = new MySqlCommand("delete from studentinfo where num=@num", con);
    cmd.Parameters.AddWithValue("@num", Convert.ToInt32(txtNum.Text));
    con.Open();
    cmd.ExecuteNonQuery();
    con.Close();
    MessageBox.Show("成功删除一条记录!");
}
```

（7）编译并运行，运行结果如图 6-9 所示。

图 6-9　StudentManageApplication 项目运行结果

6.3.3　使用 DataAdapter 对象执行数据库命令

DataAdapter 对象（当使用 MySQL 数据库时，为 MySqlDataAdapter 对象）用来连接数据源与数据集 DataSet，在一个 DataSet 对象实例中，可以包含多个 DataTable，而一个 DataTable 又可以包含多个 DataRow。DataAdapter 的主要方法描述如下。

（1）Fill()方法：用于将数据源的查询结果添加到数据集。

（2）Update()方法：用于将数据集中的数据存储到数据源中。

1．使用 DataAdapter 对象查询数据

使用 DataAdapter 对象查询数据时，首先建立数据库连接；然后利用 select 语句和数据连接建立 DataAdapter 对象，并使用 DataAdapter 对象的 Fill()方法把查询结果放在 DataSet 对象的一个数据表中；最后可使用数据控件展示数据。

2．使用 DataAdapter 对象增加、修改和删除数据

使用 DataAdapter 对象增加数据时，首先建立数据库连接；然后利用 select 语句和数据连接建立 DataAdapter 对象，建立 CommandBuilder 对象自动生成 DataAdapter 的 Command 命令，使用 DataAdapter 对象的 Fill()方法把查询结果放在 DataSet 对象的一个数据表中，实现对 DataTable 对象中数据的增加、修改或删除，通过 DataAdapter 对象的 Update()方法向数据库提交数据。

注意：只有当 DataAdapter 操作单个数据库表时，才可以利用 CommandBuilder 对象自动生成 DataAdapter 的 DeleteCommand、InsertCommand 和 UpdateCommand。为了自动生成命令，必须设置 SelectCommand 属性，SelectCommand 检索表架构以此确定自动生成的 insert、update 和 delete 语句的语法。

【例 6-4】　学生成绩管理。

【操作步骤】

（1）启动 VS，创建 Windows 窗体应用程序 StudentScoreApplication。

（2）双击 Form1.cs，切换到设计视图，从工具栏中拖曳 3 个 Label 控件、3 个 TextBox 控件、1 个 DataGridView 控件和 4 个 Button 控件到窗体设计区，并调整控件大小进行布局。

（3）在窗体设计区中右击窗体 Form1 和每一个控件，设置窗体和控件的相关属性。表 6-7 列出了新增控件属性。

表 6-7 控件属性设置

控 件	属 性	属 性 值
Form1	Text	学生成绩管理
label1	Text	学号：
textBox1	Name	txtNum
	ReadOnly	true
label2	Text	姓名：
textBox2	Name	txtName
label3	Text	成绩：
textBox3	Name	txtScore
dataGridView1	Name	dgvStudentScore
button1	Name	btnSelect
	Text	查询
button2	Name	btnInsert
	Text	增加
button3	Name	btnUpdate
	Text	修改
button4	Name	btnDelete
	Text	删除

（4）右击项目，添加应用程序配置文件 App.config，并设置连接字符串的值，程序代码如下：

```
<connectionStrings>
    <add name = "conStr" connectionString = "server=localhost;user id = root;
    password = 123456;database = studentmanage;Charset = utf8;" />
</connectionStrings>
```

（5）为项目添加引用，程序代码如下：

```
using MySql.Data.MySqlClient;
using System.Configuration;
```

（6）在 Form1 类中添加成员变量 conStr，用于保存数据库连接字符串，并创建 MySqlConnection、MySqlDataAdapter 和 DataSet 对象，程序代码如下：

```
string conStr = ConfigurationManager.ConnectionStrings["conStr"].ConnectionString;
MySqlConnection con = null;
MySqlDataAdapter da = null;
DataSet ds = null;
```

（7）双击"查询"按钮，为其添加单击事件处理程序，程序代码如下：

```
private void btnSelect_Click(object sender, EventArgs e)
{
    con = new MySqlConnection(conStr);
    da = new MySqlDataAdapter("select * from studentscore", con);
    ds = new DataSet();
    da.Fill(ds, "studentscore");
```

```
dgvStudentScore.DataSource = ds.Tables["studentscore"];
}
```

注意：DataGridView 控件是常用的数据控件，以表格的形式显示数据源中的数据。在 DataGridView 控件中进行数据绑定时，只需设置 DataSource 属性。例如，语句 "dgvStudentScore.DataSource = ds.Tables["studentscore"];" 将数据集中 studentscore 表的数据绑定到 DataGridView 控件中。

（8）双击表格 dgvStudentScore，为其添加单元格单击事件处理程序，程序代码如下：

```
private void dgvStudentScore_CellContentClick(object sender, DataGridViewCellEventArgs e)
{
    txtNum.Text = dgvStudentScore.CurrentRow.Cells["num"].Value.ToString();
    txtName.Text = dgvStudentScore.CurrentRow.Cells["name"].Value.ToString();
    txtScore.Text = dgvStudentScore.CurrentRow.Cells["score"].Value.ToString();
}
```

以上代码的功能是，当用户单击表格中的某个单元格时，将单元格中对应一行数据显示在相应的文本框中。

（9）双击"增加"按钮，为其添加单击事件处理程序，程序代码如下：

```
private void btnInsert_Click(object sender, EventArgs e)
{
    MySqlCommandBuilder builder = new MySqlCommandBuilder(da);
    DataRow r1 = ds.Tables["studentscore"].NewRow();
    r1[0] = txtNum.Text;
    r1[1] = txtName.Text;
    r1[2] = txtScore.Text;
    ds.Tables[0].Rows.Add(r1);
    da.Update(ds, "studentscore");
    dgvStudentScore.DataSource = ds.Tables["studentscore"];
}
```

（10）双击"修改"按钮，为其添加单击事件处理程序，程序代码如下：

```
private void btnUpdate_Click(object sender, EventArgs e)
{
    MySqlCommandBuilder builder = new MySqlCommandBuilder(da);
    DataRowCollection rows = ds.Tables["studentscore"].Rows;
    DataRow row;
    for (int i = 0; i < rows.Count; i++)
    {
        row = rows[i];
        if (row["num"].ToString() == txtNum.Text)
        {
            row["score"] = txtScore.Text;
        }
    }
    dgvStudentScore.DataSource = ds.Tables["studentscore"];
    da.Update(ds, "studentscore");
}
```

(11) 双击"删除"按钮，为其添加单击事件处理程序，程序代码如下：
```
private void btnDelete_Click(object sender, EventArgs e)
{
    MySqlCommandBuilder builder = new MySqlCommandBuilder(da);
    DataRowCollection rows = ds.Tables["studentscore"].Rows;
    DataRow row;
    for (int i = 0; i < rows.Count; i++)
    {
        row = rows[i];
        if (row["num"].ToString() == txtNum.Text)
        {
            row.Delete();
        }
    }
    da.Update(ds, "studentscore");
    dgvStudentScore.DataSource = ds.Tables["studentscore"];
}
```
(12) 在解决方案资源管理器中右击 StudentScoreApplication 项目，将其设为启动项目。

(13) 编译并运行，运行结果如图 6-10 所示。

图 6-10　StudentScoreApplication 项目运行结果

习题6

第 7 章 实践项目——研究生科研管理系统

本章作为课程设计的初级案例。为了提高科研工作的效率和质量，规范科研流程，优化资源配置，促进信息共享，培养研究生科研素养，并为管理层提供决策支持，从而提升研究生教育的整体水平，使用 ASP.NET 与 MySQL 数据库设计一个小型的研究生科研管理系统。

7.1 需求分析

研究生科研管理系统是一个 C/S 架构的 WinForm 项目，用户为各学院科研管理人员。主要具有导师信息管理、学生信息管理、论文成果管理、竞赛成果管理等功能。具体功能描述如下所述。

（1）研究生信息管理。用户可以对研究生信息，包括学号、姓名、年龄、性别、专业、导师姓名等信息进行增加、修改、查询和删除。

（2）导师信息管理。用户可以对研究生导师信息，包括工号、姓名、年龄、性别、职称、科研方向等信息进行增加、修改、查询和删除。

（3）论文成果管理。用户可以对研究生的论文成果信息，包括学号、姓名、作者排名、期刊名称、论文名称、发表时间等信息进行增加、修改、查询和删除。

（4）竞赛成果管理。用户可以对研究生的竞赛成果信息，包括学号、姓名、获奖排名、竞赛名称、竞赛级别、竞赛奖项、竞赛时间等信息进行增加、修改、查询和删除。

7.2 概要设计

7.2.1 架构设计

研究生科研管理系统采用三层架构设计，分为数据访问层、业务逻辑层以及表示层。

7.2.2 功能设计

根据需求分析，研究生科研管理系统功能模块设计如图 7-1 所示。

图 7-1 功能模块图

7.3 数据库设计

研究生科研管理系统采用 MySQL 数据库，名称为 srmanage，其中共 4 个数据表，分别是学生信息表、导师信息表、论文信息表、竞赛信息表。

1. 学生信息表

学生信息表用来存储研究生的相关信息，包括学号、姓名、年龄、性别、专业、导师姓名等字段，数据表的结构如表 7-1 所示。

表 7-1 学生信息表（student）

字 段 名	数 据 类 型	允 许 空 值	字 段 说 明
stuNum	int	否	学号（主键）
stuName	varchar（20）	否	姓名
stuAge	int	否	年龄
stuGender	int	否	性别
stuMajor	varchar（60）	否	专业
teaName	varchar（20）	否	导师姓名

2. 导师信息表

导师信息表用来存储导师的相关信息，包括工号、姓名、年龄、性别、职称、科研方向等字段，数据表的结构如表 7-2 所示。

表 7-2 导师信息表（teacher）

字 段 名	数 据 类 型	允 许 空 值	字 段 说 明
teaNum	int	否	工号（主键）
teaName	varchar（20）	否	姓名
teaAge	int	否	年龄
teaGender	int	否	性别
teaTitle	varchar（20）	否	职称
teaResearch	varchar（60）	否	科研方向

3. 论文信息表

论文信息表用来存储学生发表论文的相关信息，包括论文成果序号、学号、姓名、作者排名、期刊名称、论文名称、发表时间等字段，数据表的结构如表 7-3 所示。

表 7-3 论文信息表（paper）

字 段 名	数 据 类 型	允 许 空 值	字 段 说 明
paperId	int	否	论文成果序号（自增长）
stuNum	int	否	学号
stuName	varchar（20）	否	姓名
stuRanking	int	否	作者排名
journalName	varchar（100）	否	期刊名称
paperName	varchar（100）	否	论文名称
paperTime	varchar（30）	否	发表时间

4. 竞赛信息表

竞赛信息表用来存储学生竞赛获奖的相关信息，包括竞赛成果序号、学号、姓名、获奖排名、竞赛名称、竞赛级别、竞赛奖项、竞赛时间等字段，数据表的结构如表 7-4 所示。

表 7-4 竞赛信息表（competition）

字 段 名	数 据 类 型	允 许 空 值	字 段 说 明
comId	int	否	竞赛成果序号（自增长）
stuNum	int	否	学号
stuName	varchar（20）	否	姓名
stuRanking	int	否	获奖排名
comName	varchar（100）	否	竞赛名称
comLevel	varchar（20）	否	竞赛级别
comPrize	varchar（20）	否	竞赛奖项
comTime	varchar（30）	否	竞赛时间

7.4 实体模型设计

新建类库项目 Models，其中包含 4 个实体类，分别为 Student.cs、Teacher.cs、Paper.cs、Competition.cs。

1. Student 类代码

```
namespace Models
{
    public class Student
    {
        private int _stuNum;
        private string _stuName;
        private int _stuAge;
        private int _stuGender;
        private string _stuMajor;
        private string _teaName;
        public int StuNum
        {
            get { return _stuNum; }
            set { _stuNum = value; }
        }
        public string StuName
        {
            get { return _stuName; }
            set { _stuName = value; }
        }
        public int StuAge
        {
            get { return _stuAge;
```

```csharp
            set { _stuAge = value; }
        }
        public int StuGender
        {
            get { return _stuGender; }
            set { _stuGender = value; }
        }
        public string StuMajor
        {
            get { return _stuMajor; }
            set { _stuMajor = value; }
        }
        public string TeaName
        {
            get { return _teaName; }
            set { _teaName = value; }
        }
        public Student(int num, string name, int age, int gender, string major, string teacher)
        {
            _stuNum = num;
            _stuName = name;
            _stuAge = age;
            _stuGender = gender;
            _stuMajor = major;
            _teaName = teacher;
        }
        public Student() { }
    }
}
```

2. Teacher 类代码

```csharp
namespace Models
{
    public class Teacher
    {
        private int _teaNum;
        private string _teaName;
        private int _teaAge;
        private int _teaGender;
        private string _teaTitle;
        private string _teaResearch;
        public int TeaNum
        {
            get { return _teaNum; }
```

```csharp
            set { _teaNum = value; }
        }
        public string TeaName
        {
            get { return _teaName; }
            set { _teaName = value; }
        }
        public int TeaAge
        {
            get { return _teaAge; }
            set { _teaAge = value; }
        }
        public int TeaGender
        {
            get { return _teaGender; }
            set { _teaGender = value; }
        }
        public string TeaTitle
        {
            get { return _teaTitle; }
            set { _teaTitle = value; }
        }
        public string TeaResearch
        {
            get { return _teaResearch; }
            set { _teaResearch = value; }
        }
        public Teacher(int num, string name, int age, int gender, string title, string research)
        {
            _teaNum = num;
            _teaName = name;
            _teaAge = age;
            _teaGender = gender;
            _teaTitle = title;
            _teaResearch = research;
        }
        public Teacher() { }
    }
}
```

3. Paper 类代码

```csharp
namespace Models
{
    public class Paper
```

```csharp
    {
        private int _paperId;
        private int _stuNum;
        private string _stuName;
        private int _stuRanking;
        private string _journalName;
        private string _paperName;
        private string _paperTime;

        public int PaperId
        {
            get { return _paperId; }
            set { _paperId = value; }
        }
        public int StuNum
        {
            get { return _stuNum; }
            set { _stuNum = value; }
        }
        public string StuName
        {
            get { return _stuName; }
            set { _stuName = value; }
        }
        public int StuRanking
        {
            get { return _stuRanking; }
            set { _stuRanking = value; }
        }
        public string JournalName
        {
            get { return _journalName; }
            set { _journalName = value; }
        }
        public string PaperName
        {
            get { return _paperName; }
            set { _paperName = value; }
        }
        public string PaperTime
        {
            get { return _paperTime; }
            set { _paperTime = value; }
        }
    }
}
```

4. Competition 类代码

```
namespace Models
{
    public class Competition
    {
        private int _compId;
        private int _stuNum;
        private string _stuName;
        private int _stuRanking;
        private string _comName;
        private string _comLevel;
        private string _comPrize;
        private string _comTime;
        public int ComId
        {
            get { return _compId; }
            set { _compId = value; }
        }
        public int StuNum
        {
            get { return _stuNum; }
            set { _stuNum = value; }
        }
        public string StuName
        {
            get { return _stuName; }
            set { _stuName = value; }
        }
        public int StuRanking
        {
            get { return _stuRanking; }
            set { _stuRanking = value; }
        }
        public string ComName
        {
            get { return _comName; }
            set { _comName = value; }
        }
        public string ComLevel
        {
            get { return _comLevel; }
            set { _comLevel = value; }
        }
        public string ComPrize
        {
```

```
            get { return _comPrize; }
            set { _comPrize = value; }
        }
        public string ComTime
        {
            get { return _comTime; }
            set { _comTime = value; }
        }
    }
}
```

7.5 数据访问层设计

数据访问层类库 DAL 由一个核心类 DBOper.cs 组成，其中封装了执行数据的增、删、改、查等功能的方法。

1. 配置 APP.Config

```
<configuration>
  <connectionStrings>
      <add name="conStr" connectionString="server=localhost; user id=root; password=123456; database=srmanage;Charset=utf8;" />
  </connectionStrings>
</configuration>
```

2. DBOper 类代码

```
using MySql.Data.MySqlClient;
using System.Data;
using System.Configuration;
namespace DAL
{
    public class DBOper
    {
        private static MySqlConnection coon;
        public static MySqlConnection Connection
        {
            get
            {
                if (coon == null)
                {
                    string coonStr = ConfigurationManager.ConnectionStrings["conStr"].ConnectionString;
                    coon = new MySqlConnection(coonStr);
                    coon.Open();
                }
                else if (coon.State == System.Data.ConnectionState.Closed)
```

```csharp
            {
                coon.Open();
            }
            else if (coon.State == System.Data.ConnectionState.Broken)
            {
                coon.Close();
                coon.Open();
            }
            return coon;
        }
    }
    public static MySqlDataReader GetReader(string sql)
    {

        MySqlCommand cmd = new MySqlCommand(sql, Connection);
        return cmd.ExecuteReader();
    }
    public static MySqlDataReader GetReader(string sql, params MySqlParameter[] values)
    {
        MySqlCommand cmd = new MySqlCommand(sql, Connection);
        cmd.Parameters.AddRange(values);
        return cmd.ExecuteReader();
    }
    public static int ExecuteCommand(string sql)
    {
        MySqlCommand cmd = new MySqlCommand(sql, Connection);
        return cmd.ExecuteNonQuery();
    }
    public static int ExecuteCommand(string sql, params MySqlParameter[] values)
    {
        MySqlCommand cmd = new MySqlCommand(sql, Connection);
        cmd.Parameters.AddRange(values);
        return cmd.ExecuteNonQuery();
    }
    public static DataTable GetDataTable(string sql)
    {
        MySqlDataAdapter da = new MySqlDataAdapter(sql, Connection);
        DataSet ds = new DataSet();
        da.Fill(ds);
        return ds.Tables[0];
    }
    public static DataTable GetDataTable(string sql, params MySqlParameter[] values)
```

```
            {
                MySqlCommand cmd = new MySqlCommand(sql, Connection);
                cmd.Parameters.AddRange(values);
                MySqlDataAdapter da = new MySqlDataAdapter(cmd);
                DataSet ds = new DataSet();
                da.Fill(ds);
                return ds.Tables[0];
            }
        }
    }
```

7.6 业务逻辑层设计

业务逻辑层类库 BLL 由 4 个核心类组成,分别为 StudentManager.cs、TeacherManager.cs、PaperManager.cs、CompetitionManager.cs。StudentManager 类用于封装学生相关功能方法;TeacherManager 类用于封装导师相关功能方法;PaperManager 类用于封装论文成果相关功能方法;CompetitionManager 类用于封装竞赛成果相关功能方法。

1. StudentManager 类代码

```
using System;
using Models;
using MySql.Data.MySqlClient;
using DAL;
using System.Data;
namespace BLL
{
    public class StudentManager
    {
        //(1)将学生信息保存到数据库student表中
        public static bool AddStudent(Student student)
        {
            string sql = "insert into student values(@stuNum,@stuName,@stuAge,@stuGender,@stuMajor,@teaName)";
            MySqlParameter p1 = new MySqlParameter("@stuNum", student.StuNum);
            MySqlParameter p2 = new MySqlParameter("@stuName", student.StuName);
            MySqlParameter p3 = new MySqlParameter("@stuAge", student.StuAge);
            MySqlParameter p4 = new MySqlParameter("@stuGender", student.StuGender);
            MySqlParameter p5 = new MySqlParameter("@stuMajor", student.StuMajor);
            MySqlParameter p6 = new MySqlParameter("@teaName", student.TeaName);
            if (DBOper.ExecuteCommand(sql, p1, p2, p3, p4, p5, p6) == 1)
            {
```

```csharp
            return true;
        }
        else
        {
            return false;
        }
    }
    //(2)根据学生姓名进行模糊查询全部学生信息
    public static DataTable GetALLStudent(string name)
    {
        string sql = @"select stuNum as 学号,stuName as 姓名,stuMajor as 专业,teaName as 导师 from student where stuName like '%" + name + "%'";
        DataTable dt = DBOper.GetDataTable(sql);
        return dt;
    }
    //(3)根据学生学号查询该学生信息
    public static Student GetStudent(int num)
    {
        Student student = null;
        string sql = "select * from student where stuNum=\'" + num + "\'";
        MySqlDataReader reader = DBOper.GetReader(sql);
        if (reader.Read())
        {
            student = new Student();
            student.StuNum = Convert.ToInt32(reader["stuNum"]);
            student.StuName = Convert.ToString(reader["stuName"]);
            student.StuAge = Convert.ToInt32(reader["stuAge"]);
            student.StuGender = Convert.ToInt32(reader["stuGender"]);
            student.StuMajor = Convert.ToString(reader["stuMajor"]);
            student.TeaName = Convert.ToString(reader["teaName"]);
        }
        reader.Close();
        return student;
    }
    //(4)修改学生信息
    public static bool UpdateStudent(Student student)
    {
        string sqlUpdateUser = @"update student set stuName=@stuName,stuAge=@stuAge,stuGender=@stuGender,stuMajor=@stuMajor,teaName=@teaName where stuNum=@stuNum";
        MySqlParameter p1 = new MySqlParameter("@stuName", student.StuName);
        MySqlParameter p2 = new MySqlParameter("@stuAge", student.StuAge);
        MySqlParameter p3 = new MySqlParameter("@stuGender", student.StuGender);
```

```csharp
            MySqlParameter p4 = new MySqlParameter("@stuMajor", 
student.StuMajor);
            MySqlParameter p5 = new MySqlParameter("@teaName", 
student.TeaName);
            MySqlParameter p6 = new MySqlParameter("@stuNum", student.StuNum);
            MySqlParameter[] paramArray = new MySqlParameter[] { p1, p2, p3, 
p4, p5,p6 };
            if (DBOper.ExecuteCommand(sqlUpdateUser, paramArray) == 1)
            {
                return true;
            }
            else
            {
                return false;
            }
        }
        //(5)删除学生信息
        public static bool DeleteStudent(int stuNum)
        {
            string sqlDeleteUser = @"delete from student where stuNum=@stuNum";
            MySqlParameter p1 = new MySqlParameter("@stuNum", stuNum);
            MySqlParameter[] paramArray = new MySqlParameter[] { p1 };
            if (DBOper.ExecuteCommand(sqlDeleteUser, paramArray) == 1)
            {
                return true;
            }
            else
            {
                return false;
            }
        }
        //(6)根据导师姓名进行模糊查询全部学生信息
        public static DataTable GetALLStudentByTeaName(string name)
        {
            string sql = @"select stuNum as 学号,stuName as 姓名,stuMajor as 
专业,teaName as 导师 from student where teaName like '%" + name + "%'";
            DataTable dt = DBOper.GetDataTable(sql);
            return dt;
        }
    }
}
```

2. TeacherManager 类代码

```csharp
using System;
using Models;
```

```csharp
using MySql.Data.MySqlClient;
using DAL;
using System.Data;
namespace BLL
{
    public class TeacherManager
    {
        //(1)从teacher表中查询所有的导师姓名
        public static DataTable GetAllTeacher()
        {
            string sql = @"select teaName from teacher";
            DataTable dt = DBOper.GetDataTable(sql);
            return dt;
        }
        //(2)将导师信息保存到数据库teacher表中
        public static bool AddTeacher(Teacher teacher)
        {
            string sql = "insert into teacher values(@teaNum,@teaName,@teaAge,@teaGender,@teaTitle,@teaResearch)";
            MySqlParameter p1 = new MySqlParameter("@teaNum", teacher.TeaNum);
            MySqlParameter p2 = new MySqlParameter("@teaName", teacher.TeaName);
            MySqlParameter p3 = new MySqlParameter("@teaAge", teacher.TeaAge);
            MySqlParameter p4 = new MySqlParameter("@teaGender", teacher.TeaGender);
            MySqlParameter p5 = new MySqlParameter("@teaTitle", teacher.TeaTitle);
            MySqlParameter p6 = new MySqlParameter("@teaResearch", teacher.TeaResearch);
            if (DBOper.ExecuteCommand(sql, p1, p2, p3, p4, p5, p6) == 1)
            {
                return true;
            }
            else
            {
                return false;
            }
        }
        //(3)根据导师姓名进行模糊查询全部导师信息
        public static DataTable GetALLTeacher(string name)
        {
            string sql = @"select teaNum as 工号,teaName as 姓名,teaTitle as 职称,teaResearch as 科研方向 from teacher where teaName like '%" + name + "%'";
            DataTable dt = DBOper.GetDataTable(sql);
            return dt;
        }
```

```csharp
//(4)根据导师工号查询该导师信息
public static Teacher GetTeacher(int num)
{
    Teacher teacher = null;
    string sql = "select * from teacher where teaNum=\'" + num + "\'";
    MySqlDataReader reader = DBOper.GetReader(sql);
    if (reader.Read())
    {
        teacher = new Teacher();
        teacher.TeaNum = Convert.ToInt32(reader["teaNum"]);
        teacher.TeaName = Convert.ToString(reader["teaName"]);
        teacher.TeaAge = Convert.ToInt32(reader["teaAge"]);
        teacher.TeaGender = Convert.ToInt32(reader["teaGender"]);
        teacher.TeaTitle = Convert.ToString(reader["teaTitle"]);
        teacher.TeaResearch = Convert.ToString(reader["teaResearch"]);
    }
    reader.Close();
    return teacher;
}
//(5)修改导师信息
public static bool UpdateTeacher(Teacher teacher)
{
    string sqlUpdateUser = @"update teacher set teaName=@teaName,teaAge=@teaAge,teaGender=@teaGender,teaTitle=@teaTitle,teaResearch=@teaResearch where teaNum=@teaNum";
    MySqlParameter p1 = new MySqlParameter("@teaName", teacher.TeaName);
    MySqlParameter p2 = new MySqlParameter("@teaAge", teacher.TeaAge);
    MySqlParameter p3 = new MySqlParameter("@teaGender", teacher.TeaGender);
    MySqlParameter p4 = new MySqlParameter("@teaTitle", teacher.TeaTitle);
    MySqlParameter p5 = new MySqlParameter("@teaResearch", teacher.TeaResearch);
    MySqlParameter p6 = new MySqlParameter("@teaNum", teacher.TeaNum);
    MySqlParameter[] paramArray = new MySqlParameter[] { p1, p2, p3, p4, p5, p6 };
    if (DBOper.ExecuteCommand(sqlUpdateUser, paramArray) == 1)
    {
        return true;
    }
    else
    {
        return false;
    }
}
```

```csharp
//(6)删除导师信息
public static bool DeleteTeacher(int teaNum)
{
    string sqlDeleteUser = @"delete from teacher where teaNum=@teaNum";
    MySqlParameter p1 = new MySqlParameter("@teaNum", teaNum);
    MySqlParameter[] paramArray = new MySqlParameter[] { p1 };
    if (DBOper.ExecuteCommand(sqlDeleteUser, paramArray) == 1)
    {
        return true;
    }
    else
    {
        return false;
    }
}
}
```

3. PaperManager 类代码

```csharp
using Models;
using MySql.Data.MySqlClient;
using DAL;
using System.Data;
namespace BLL
{
    public class PaperManager
    {
        //(1)将论文成果信息保存到数据库paper表中
        public static bool AddPaper(Paper paper)
        {
            string sql = "insert into paper values(null,@stuNum,@stuName,@stuRanking,@journalName,@paperName,@paperTime)";
            MySqlParameter p1 = new MySqlParameter("@stuNum", paper.StuNum);
            MySqlParameter p2 = new MySqlParameter("@stuName", paper.StuName);
            MySqlParameter p3 = new MySqlParameter("@stuRanking", paper.StuRanking);
            MySqlParameter p4 = new MySqlParameter("@journalName", paper.JournalName);
            MySqlParameter p5 = new MySqlParameter("@paperName", paper.PaperName);
            MySqlParameter p6 = new MySqlParameter("@paperTime", paper.PaperTime);
            if (DBOper.ExecuteCommand(sql, p1, p2, p3, p4, p5, p6) == 1)
            {
```

```csharp
            return true;
        }
        else
        {
            return false;
        }
    }
    //(2)根据学生姓名进行查询全部论文成果信息
    public static DataTable GetALLPaper(string name)
    {
        string sql = @"select paperId as ID,stuNum as 学号,stuName as 姓名,stuRanking as 作者排名,journalName as 期刊名称,paperName as 论文名称,paperTime as 发表时间 from paper where stuName=@name";
        MySqlParameter p1 = new MySqlParameter("@name", name);
        DataTable dt = DBOper.GetDataTable(sql, p1);
        return dt;
    }
    //(3)修改学生论文成果信息
    public static bool UpdatePaper(Paper paper)
    {
        string sqlUpdateUser = @"update paper set stuRanking=@stuRanking,journalName=@journalName,paperName=@paperName,paperTime=@paperTime where paperId=@paperId";
        MySqlParameter p1 = new MySqlParameter("@stuRanking", paper.StuRanking);
        MySqlParameter p2 = new MySqlParameter("@journalName", paper.JournalName);
        MySqlParameter p3 = new MySqlParameter("@paperName", paper.PaperName);
        MySqlParameter p4 = new MySqlParameter("@paperTime", paper.PaperTime);
        MySqlParameter p5 = new MySqlParameter("@paperId", paper.PaperId);
        MySqlParameter[] paramArray = new MySqlParameter[] { p1, p2, p3, p4, p5 };
        if (DBOper.ExecuteCommand(sqlUpdateUser, paramArray) == 1)
        {
            return true;
        }
        else
        {
            return false;
        }
    }
    //(4)删除学生论文成果信息
```

```csharp
        public static bool DeletePaper(int paperId)
        {
            string sqlDeleteUser = @"delete from paper where paperId=@paperId";
            MySqlParameter p1 = new MySqlParameter("@paperId", paperId);
            MySqlParameter[] paramArray = new MySqlParameter[] { p1 };
            if (DBOper.ExecuteCommand(sqlDeleteUser, paramArray) == 1)
            {
                return true;
            }
            else
            {
                return false;
            }
        }
        //(5)根据学生学号进行查询该学生全部论文成果信息
        public static DataTable GetALLPaperByNum(int num)
        {
            string sql = @"select stuNum as 学号,stuName as 姓名,stuRanking as 作者排名,journalName as 期刊名称,paperName as 论文名称,paperTime as 发表时间 from paper where stuNum=@stuNum";
            MySqlParameter p1 = new MySqlParameter("@stuNum", num);
            DataTable dt = DBOper.GetDataTable(sql, p1);
            return dt;
        }
    }
}
```

4. CompetitionManager 类代码

```csharp
using Models;
using MySql.Data.MySqlClient;
using DAL;
using System.Data;
namespace BLL
{
    public class CompetitionManager
    {
        //(1)将竞赛成果信息保存到数据库competition表中
        public static bool AddCompetition(Competition competition)
        {
            string sql = "insert into competition values(null,@stuNum,@stuName,@stuRanking,@comName,@comLevel,@comPrize,@comTime)";
            MySqlParameter p1 = new MySqlParameter("@stuNum", competition.StuNum);
            MySqlParameter p2 = new MySqlParameter("@stuName", competition.StuName);
```

```csharp
            MySqlParameter p3 = new MySqlParameter("@stuRanking", competition.StuRanking);
            MySqlParameter p4 = new MySqlParameter("@comName", competition.ComName);
            MySqlParameter p5 = new MySqlParameter("@comLevel", competition.ComLevel);
            MySqlParameter p6 = new MySqlParameter("@comPrize", competition.ComPrize);
            MySqlParameter p7 = new MySqlParameter("@comTime", competition.ComTime);
            if (DBOper.ExecuteCommand(sql, p1, p2, p3, p4, p5, p6, p7) == 1)
            {
                return true;
            }
            else
            {
                return false;
            }
        }
        //(2)根据学生姓名进行查询全部竞赛成果信息
        public static DataTable GetALLCompetition(string name)
        {
            string sql = @"select comId as ID,stuNum as 学号,stuName as 姓名,stuRanking as 获奖排名,comName as 竞赛名称,comLevel as 竞赛级别,comPrize as 竞赛奖项,comTime as 竞赛时间 from competition where stuName=@name";
            MySqlParameter p1 = new MySqlParameter("@name", name);
            DataTable dt = DBOper.GetDataTable(sql, p1);
            return dt;
        }
        //(3)修改学生竞赛成果信息
        public static bool UpdateCompetition(Competition competition)
        {
            string sqlUpdateUser = @"update competition set stuRanking = @stuRanking,comName = @comName,comLevel = @comLevel,comPrize = @comPrize,comTime = @comTime where comId = @comId";
            MySqlParameter p1 = new MySqlParameter("@stuRanking", competition.StuRanking);
            MySqlParameter p2 = new MySqlParameter("@comName", competition.ComName);
            MySqlParameter p3 = new MySqlParameter("@comLevel", competition.ComLevel);
            MySqlParameter p4 = new MySqlParameter("@comPrize", competition.ComPrize);
            MySqlParameter p5 = new MySqlParameter("@comTime", competition.ComTime);
```

```csharp
            MySqlParameter p6 = new MySqlParameter("@comId",
competition.ComId);
            MySqlParameter[] paramArray = new MySqlParameter[] { p1, p2,
p3, p4, p5, p6 };
            if (DBOper.ExecuteCommand(sqlUpdateUser, paramArray) == 1)
            {
                return true;
            }
            else
            {
                return false;
            }
        }
        //(4)删除学生竞赛成果信息
        public static bool DeleteCompetition(int comId)
        {
            string sqlDeleteUser = @"delete from competition where comId = @comId";
            MySqlParameter p1 = new MySqlParameter("@comId", comId);
            MySqlParameter[] paramArray = new MySqlParameter[] { p1 };
            if (DBOper.ExecuteCommand(sqlDeleteUser, paramArray) == 1)
            {
                return true;
            }
            else
            {
                return false;
            }
        }
        //(5)根据学生学号进行查询该学生全部竞赛成果信息
        public static DataTable GetALLCompetitionByNum(int num)
        {
            string sql = @"select stuNum as 学号,stuName as 姓名,stuRanking as 获奖排名,comName as 竞赛名称,comLevel as 竞赛级别,comPrize as 竞赛奖项,comTime as 竞赛时间 from competition where stuNum = @stuNum";
            MySqlParameter p1 = new MySqlParameter("@stuNum", num);
            DataTable dt = DBOper.GetDataTable(sql, p1);
            return dt;
        }
    }
}
```

7.7 表示层设计

研究生科研管理系统中共包含 11 个窗体，具体窗体信息描述如表 7-5 所示。

表 7-5 研究生科研管理系统窗体描述表

序号	模块	窗体 Text 属性值	窗体 Name 属性值
1	系统主窗体	研究生科研管理系统	MainForm
2	研究生信息管理	添加学生信息	AddStudent
3		维护学生信息	MaintainStudent
4	导师信息管理	添加导师信息	AddTeacher
5		维护导师信息	MaintainTeacher
6	论文成果管理	添加论文成果	AddPaper
7		维护论文成果	MaintainPaper
8	竞赛成果管理	添加竞赛成果	AddCompetition
9		维护竞赛成果	MaintainCompetition
10	成果查询	查询导师成果	TeacherSearch
11		查询学生成果	StudentSearch

7.7.1 系统主窗体设计

1. 窗体设计

系统主窗体如图 7-2 所示。

图 7-2 系统主窗体

系统主窗体和主要控件的属性设置如表 7-6 所示。

表 7-6 系统主窗体和主要控件的属性设置

Name	类型	属性
MainForm	Form	WindowState：Maximized IsMdiContainer：True
AddStudent	ToolStripMenuItem	Text：添加学生信息
MaintainStudent	ToolStripMenuItem	Text：维护学生信息
AddTeacher	ToolStripMenuItem	Text：添加导师信息
MaintainTeacher	ToolStripMenuItem	Text：维护导师信息
AddPaper	ToolStripMenuItem	Text：添加论文成果
MaintainPaper	ToolStripMenuItem	Text：维护论文成果
AddCompetition	ToolStripMenuItem	Text：添加竞赛成果
MaintainCompetition	ToolStripMenuItem	Text：维护竞赛成果
TeacherSearch	ToolStripMenuItem	Text：查询导师成果
StudentSearch	ToolStripMenuItem	Text：查询学生成果

2. 功能实现代码

```
using System;
using System.Windows.Forms;
namespace StudentAchievementManagementSystem
{
    public partial class MainForm : Form
    {
        public MainForm()
        {
            InitializeComponent();
        }
        private void AddStudent_Click(object sender, EventArgs e)
        {
            AddStudent  f = new AddStudent();
            f.MdiParent = this;
            f.Show();
        }
        private void MaintainStudent_Click(object sender, EventArgs e)
        {
            MaintainStudent f = new MaintainStudent();
            f.MdiParent = this;
            f.Show();
        }
        private void AddTeacher_Click(object sender, EventArgs e)
        {
            AddTeacher f = new AddTeacher();
            f.MdiParent = this;
            f.Show();
        }
```

```csharp
private void MaintainTeacher_Click(object sender, EventArgs e)
{
    MaintainTeacher f = new MaintainTeacher();
    f.MdiParent = this;
    f.Show();
}
private void AddPaper_Click(object sender, EventArgs e)
{
    AddPaper f = new AddPaper();
    f.MdiParent = this;
    f.Show();
}
private void MaintainPaper_Click(object sender, EventArgs e)
{
    MaintainPaper f = new MaintainPaper();
    f.MdiParent = this;
    f.Show();
}
private void AddCompetition_Click(object sender, EventArgs e)
{
    AddCompetition f = new AddCompetition();
    f.MdiParent = this;
    f.Show();
}
private void MaintainCompetition_Click(object sender, EventArgs e)
{
    MaintainCompetition f = new MaintainCompetition();
    f.MdiParent = this;
    f.Show();
}
private void TeacherSearch_Click(object sender, EventArgs e)
{
    TeacherSearch f = new TeacherSearch();
    f.MdiParent = this;
    f.Show();
}
private void StudentSearch_Click(object sender, EventArgs e)
{
    StudentSearch f = new StudentSearch();
    f.MdiParent = this;
    f.Show();
}
    }
}
```

7.7.2 添加学生信息设计

1. 窗体设计

"添加学生信息"窗体如图7-3所示。

图7-3 "添加学生信息"窗体

"添加学生信息"窗体和主要控件的属性设置如表7-7所示。

表7-7 "添加学生信息"窗体和主要控件的属性设置

Name	类型	Text属性	其他属性
AddStudent	Form	添加学生信息	
groupBox1	GroupBox	学生信息	
txtNumber	TextBox		
txtName	TextBox		
txtAge	TextBox		
rbMale	RadioButton	男	Checked：True
rbFemale	RadioButton	女	
cmbTeacher	ComboBox		
cmbMajor	ComboBox		
btnAdd	Button	添加	
btnExit	Button	退出	

2. 功能实现代码

```
using System;
using System.Data;
using System.Windows.Forms;
using Models;
using BLL;
namespace StudentAchievementManagementSystem
{
```

```csharp
public partial class AddStudent : Form
{
    public AddStudent()
    {
        InitializeComponent();
    }
    private void AddStudent_Load(object sender, EventArgs e)
    {
        //专业下拉列表框加载值
        cmbMajor.Items.Add("计算机应用技术");
        cmbMajor.Items.Add("软件工程");
        //导师下拉列表框加载值
        DataTable dt = TeacherManager.GetAllTeacher();
        if (dt.Rows.Count > 0)
        {
            for (int i = 0; i <= dt.Rows.Count - 1; i++)
            {
                cmbTeacher.Items.Add(dt.Rows[i]["teaName"].ToString());
            }
        }
    }
    private void btnAdd_Click(object sender, EventArgs e)
    {
        //1.接收用户输入的数据
        int num = Convert.ToInt32(txtNumber.Text);
        string name = txtName.Text;
        int age = Convert.ToInt32(txtAge.Text);
        int gender = 0;
        if (rbFemale.Checked)
        {
            gender = 1;
        }
        string major = cmbMajor.Text;
        string teacher = cmbTeacher.Text;
        // 2.存入数据库表
        Student student = new Student(num, name, age, gender, major, teacher);
        if (StudentManager.AddStudent(student))
        {
            MessageBox.Show("添加学生信息成功！");
        }
        else
        {
            MessageBox.Show("添加学生信息失败！");
        }
```

```
        }
        private void btnExit_Click(object sender, EventArgs e)
        {
            this.Close();
        }
    }
}
```

7.7.3 维护学生信息设计

1. 窗体设计

"维护学生信息"窗体如图 7-4 所示。

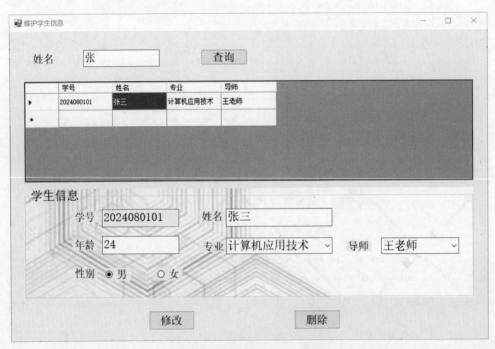

图 7-4 "维护学生信息"窗体

"维护学生信息"窗体和主要控件的属性设置如表 7-8 所示。

表 7-8 "维护学生信息"窗体和控件的属性设置

Name	类　　型	Text 属性	其 他 属 性
MaintainStudent	Form	维护学生信息	
txtQueryName	TextBox		
btnQuery	Button	查询	
dgvStudent	DataGridView		
groupBox1	GroupBox	学生信息	
txtNumber	TextBox		ReadOnly：True
txtName	TextBox		

续表

Name	类型	Text 属性	其他属性
txtAge	TextBox		
cmbMajor	ComboBox		
cmbTeacher	ComboBox		
rbMale	RadioButton	男	Checked：True
rbFemale	RadioButton	女	
btnUpdate	Button	修改	
btnDelete	Button	删除	

2. 功能实现代码

```csharp
using System;
using System.Data;
using System.Windows.Forms;
using Models;
using BLL;
namespace StudentAchievementManagementSystem
{
    public partial class MaintainStudent : Form
    {
        public MaintainStudent()
        {
            InitializeComponent();
        }
        private void btnQuery_Click(object sender, EventArgs e)
        {
            //根据用户输入学生姓名模糊查询并展示查询结果
            string name = txtQueryName.Text;
            DataTable dt = StudentManager.GetALLStudent(name);
            dgvStudent.DataSource = dt;
        }
        private void dgvStudent_CellContentClick(object sender, DataGridViewCellEventArgs e)
        {
            int num = Convert.ToInt32(dgvStudent.CurrentRow.Cells["学号"].Value);
            Student student = StudentManager.GetStudent(num);
            txtNumber.Text = student.StuNum.ToString();
            txtName.Text = student.StuName;
            txtAge.Text = student.StuAge.ToString();
            if (student.StuGender == 1)
            {
                rbFemale.Checked = true;
            }
```

```csharp
            else
            {
                rbMale.Checked = true;
            }
            cmbMajor.Text = student.StuMajor;
            cmbTeacher.Text = student.TeaName;
        }
        private void btnUpdate_Click(object sender, EventArgs e)
        {
            //创建学生对象,并赋值
            Student student = new Student();
            student.StuNum = Convert.ToInt32(txtNumber.Text);
            student.StuName = txtName.Text;
            student.StuAge = Convert.ToInt32(txtAge.Text);
            int gender = 0;
            if (rbFemale.Checked)
            {
                gender = 1;
            }
            student.StuGender = gender;
            student.StuMajor = cmbMajor.Text;
            student.TeaName = cmbTeacher.Text;
            //修改学生信息
            if (StudentManager.UpdateStudent(student))
            {
                MessageBox.Show("学生信息修改成功!");
                //更新展示查询结果
                DataTable dt = StudentManager.GetALLStudent(txtQueryName.Text);
                dgvStudent.DataSource = dt;
            }
            else
            {
                MessageBox.Show("学生信息修改失败!");
            }
        }
        private void btnDelete_Click(object sender, EventArgs e)
        {
            if (MessageBox.Show("确定要删除学生信息吗?", "提示",
MessageBoxButtons.OKCancel, MessageBoxIcon.Information) == DialogResult.OK)
            {
                //删除学生信息
                if (StudentManager.DeleteStudent(Convert.ToInt32
(txtNumber.Text)))
                {
                    MessageBox.Show("学生信息删除成功!");
```

```
                    //更新展示查询结果
                    DataTable dt = StudentManager.GetALLStudent
(txtQueryName.Text);
                    dgvStudent.DataSource = dt;
                }
                else
                {
                    MessageBox.Show("学生信息删除失败！");
                }
            }
        }
    }
}
```

7.7.4 添加导师信息设计

1. 窗体设计

"添加导师信息"窗体如图 7-5 所示。

图 7-5 "添加导师信息"窗体

"添加导师信息"窗体和控件的属性设置如表 7-9 所示。

表 7-9 "添加导师信息"窗体和控件的属性设置

Name	类型	Text 属性	其他属性
AddTeacher	Form	添加导师信息	
groupBox1	GroupBox	导师信息	
txtNumber	TextBox		

Name	类型	Text 属性	其他属性
txtName	TextBox		
txtAge	TextBox		
rbMale	RadioButton	男	Checked：True
rbFemale	RadioButton	女	
cmbTitle	ComboBox		
txtResearch	TextBox		Multiline：True
btnAdd	Button	添加	
btnExit	Button	退出	

2. 功能实现代码

```
using System;
using System.Windows.Forms;
using Models;
using BLL;
namespace StudentAchievementManagementSystem
{
    public partial class AddTeacher : Form
    {
        public AddTeacher()
        {
            InitializeComponent();
        }
        private void AddTeacher_Load(object sender, EventArgs e)
        {
            //导师职称下拉列表框加载值
            cmbTitle.Items.Add("教授");
            cmbTitle.Items.Add("副教授");
            cmbTitle.Items.Add("讲师");
            cmbTitle.Items.Add("助教");
        }
        private void btnAdd_Click(object sender, EventArgs e)
        {
            //1.接收用户输入的数据
            int num = Convert.ToInt32(txtNumber.Text);
            string name = txtName.Text;
            int age = Convert.ToInt32(txtAge.Text);
            int gender = 0;
            if (rbFemale.Checked)
            {
                gender = 1;
            }
            string title = cmbTitle.Text;
```

```
            string research = txtResearch.Text;
            // 2.存入数据库表
            Teacher teacher = new Teacher(num, name, age, gender, title, research);
            if (TeacherManager.AddTeacher(teacher))
            {
                MessageBox.Show("添加导师信息成功！");
            }
            else
            {
                MessageBox.Show("添加导师信息失败！");
            }
        }
        private void btnExit_Click(object sender, EventArgs e)
        {
            this.Close();
        }
    }
}
```

7.7.5 维护导师信息设计

1. 窗体设计

"维护导师信息"窗体如图 7-6 所示。

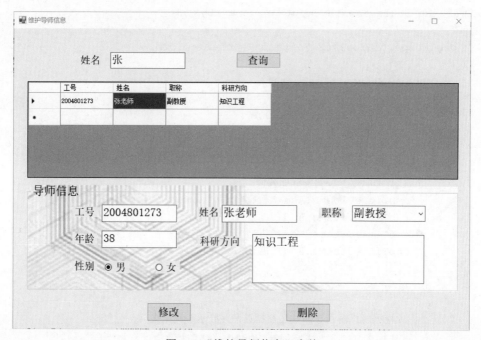

图 7-6 "维护导师信息"窗体

"维护导师信息"窗体和控件的属性设置如表 7-10 所示。

表 7-10 "维护导师信息"窗体和控件的属性设置

Name	类 型	Text 属性	其 他 属 性
MaintainTeacher	Form	维护导师信息	
txtQueryName	GroupBox		
btnQuery	Button	查询	
dgvTeacher	DataGridView		
groupBox1	GroupBox	导师信息	
txtNumber	TextBox		ReadOnly：True
txtName	TextBox		
txtAge	TextBox		
txtResearch	TextBox		Multiline：True
rbMale	RadioButton	男	Checked：True
rbFemale	RadioButton	女	
cmbTitle	ComboBox		
btnUpdate	Button	修改	
btnDelete	Button	删除	

2. 功能实现代码

```csharp
using System;
using System.Data;
using System.Windows.Forms;
using Models;
using BLL;
namespace StudentAchievementManagementSystem
{
    public partial class MaintainTeacher : Form
    {
        public MaintainTeacher()
        {
            InitializeComponent();
        }
        private void MaintainTeacher_Load(object sender, EventArgs e)
        {
            //导师职称下拉列表框加载值
            cmbTitle.Items.Add("教授");
            cmbTitle.Items.Add("副教授");
            cmbTitle.Items.Add("讲师");
            cmbTitle.Items.Add("助教");
        }
        private void btnQuery_Click(object sender, EventArgs e)
        {
            //根据用户输入导师姓名模糊查询并展示查询结果
            string name = txtQueryName.Text;
            DataTable dt = TeacherManager.GetALLTeacher(name);
            dgvTeacher.DataSource = dt;
        }
```

```csharp
        private void dgvTeacher_CellContentClick(object sender,
DataGridViewCellEventArgs e)
        {
            int num = Convert.ToInt32(dgvTeacher.CurrentRow
.Cells["工号"].Value);
            Teacher teacher = TeacherManager.GetTeacher(num);
            txtNumber.Text = teacher.TeaNum.ToString();
            txtName.Text = teacher.TeaName;
            txtAge.Text = teacher.TeaAge.ToString();
            if (teacher.TeaGender == 1)
            {
                rbFemale.Checked = true;
            }
            else
            {
                rbMale.Checked = true;
            }
            cmbTitle.Text = teacher.TeaTitle;
            txtResearch.Text = teacher.TeaResearch;
        }
        private void btnUpdate_Click(object sender, EventArgs e)
        {
            //创建导师对象,并赋值
            Teacher teacher = new Teacher();
            teacher.TeaNum = Convert.ToInt32(txtNumber.Text);
            teacher.TeaName = txtName.Text;
            teacher.TeaAge = Convert.ToInt32(txtAge.Text);
            int gender = 0;
            if (rbFemale.Checked)
            {
                gender = 1;
            }
            teacher.TeaGender = gender;
            teacher.TeaTitle = cmbTitle.Text;
            teacher.TeaResearch = txtResearch.Text;
            //修改导师信息
            if (TeacherManager.UpdateTeacher(teacher))
            {
                MessageBox.Show("导师信息修改成功!");
                //更新展示查询结果
                DataTable dt = TeacherManager.GetALLTeacher(txtQueryName.Text);
                dgvTeacher.DataSource = dt;
            }
            else
            {
                MessageBox.Show("导师信息修改失败!");
            }
        }
```

```csharp
private void btnDelete_Click(object sender, EventArgs e)
{
    if (MessageBox.Show("确定要删除导师信息吗？", "提示", MessageBoxButtons.OKCancel, MessageBoxIcon.Information) == DialogResult.OK)
    {
        //删除导师信息
        if (TeacherManager.DeleteTeacher(Convert.ToInt32(txtNumber.Text)))
        {
            MessageBox.Show("导师信息删除成功！");
            //更新展示查询结果
            DataTable dt = TeacherManager.GetALLTeacher(txtQueryName.Text);
            dgvTeacher.DataSource = dt;
        }
        else
        {
            MessageBox.Show("导师信息删除失败！");
        }
    }
}
```

7.7.6 添加论文成果设计

1. 窗体设计

"添加论文成果"窗体如图 7-7 所示。

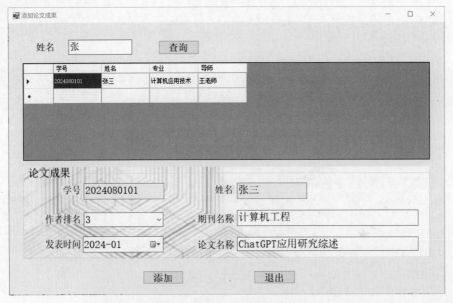

图 7-7 "添加论文成果"窗体

"添加论文成果"窗体和主要控件的属性设置如表 7-11 所示。

表 7-11 "添加论文成果"窗体和主要控件的属性设置

Name	类 型	Text 属性	其 他 属 性
AddPaper	Form	添加论文成果	
txtQueryName	TextBox		
btnQuery	Button	查询	
dgvStudent	DataGridView		
groupBox1	GroupBox	论文成果	
txtNumber	TextBox		ReadOnly:True
txtName	TextBox		ReadOnly:True
cmbRanking	ComboBox		
txtJournalName	TextBox		
txtPaperName	TextBox		
dtpTime	DateTimePicker		
btnAdd	Button	添加	
btnExit	Button	退出	

2. 功能实现代码

```
using System;
using System.Data;
using System.Windows.Forms;
using BLL;
using Models;

namespace StudentAchievementManagementSystem
{
    public partial class AddPaper : Form
    {
        public AddPaper()
        {
            InitializeComponent();
        }
        private void AddPaper_Load(object sender, EventArgs e)
        {
            cmbRanking.Items.Add("1");
            cmbRanking.Items.Add("2");
            cmbRanking.Items.Add("3");
            cmbRanking.Items.Add("4");
            cmbRanking.Items.Add("5");
            cmbRanking.Items.Add("6及以后");
        }
        private void btnQuery_Click(object sender, EventArgs e)
        {
```

```csharp
            //根据用户输入学生姓名模糊查询并展示查询结果
            string name = txtQueryName.Text;
            DataTable dt = StudentManager.GetALLStudent(name);
            dgvStudent.DataSource = dt;
        }
        private void dgvStudent_CellContentClick(object sender, DataGridViewCellEventArgs e)
        {
            //用户单击数据后为学号和姓名文本框赋值
            txtNumber.Text = dgvStudent.CurrentRow.Cells["学号"].Value.ToString();
            txtName.Text = dgvStudent.CurrentRow.Cells["姓名"].Value.ToString();
        }
        private void btnAdd_Click(object sender, EventArgs e)
        {
            //1.创建论文成果对象，并赋值
            Paper paper = new Paper();
            paper.StuNum = Convert.ToInt32(txtNumber.Text);
            paper.StuName = txtName.Text;
            paper.StuRanking = Convert.ToInt32(cmbRanking.Text);
            paper.JournalName = txtJournalName.Text;
            paper.PaperName = txtPaperName.Text;
            paper.PaperTime= dtpTime.Text;
            // 2.存入数据库表
            if (PaperManager.AddPaper(paper))
            {
                MessageBox.Show("添加论文成果成功！");
            }
            else
            {
                MessageBox.Show("添加论文成果失败！");
            }
        }
        private void btnExit_Click(object sender, EventArgs e)
        {
            this.Close();
        }
    }
}
```

7.7.7 维护论文成果设计

1. 窗体设计

"维护论文成果"窗体如图 7-8 所示。

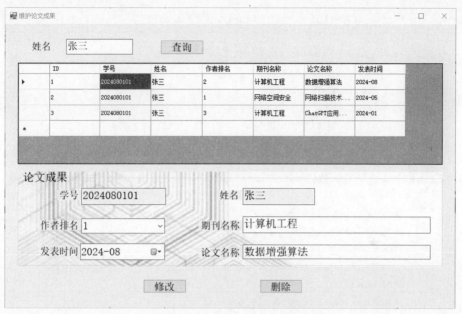

图 7-8 "维护论文成果"窗体

"维护论文成果"窗体和主要控件的属性设置如表 7-12 所示。

表 7-12 "维护论文成果"窗体和主要控件的属性设置

Name	类 型	Text 属性	其 他 属 性
MaintainPaper	Form	维护论文成果	
txtQueryName	TextBox		
btnQuery	Button	查询	
dgvPaper	DataGridView		
groupBox1	GroupBox	论文成果	
txtNumber	TextBox		ReadOnly：True
txtName	TextBox		ReadOnly：True
cmbRanking	ComboBox		
txtJournalName	TextBox		
txtPaperName	TextBox		
txtJournalName	TextBox		
dtpTime	DateTimePicker		
btnUpdate	Button	修改	
btnDelete	Button	删除	

2. 功能实现代码

```
using System;
using System.Data;
using System.Windows.Forms;
using BLL;
using Models;
```

```csharp
namespace StudentAchievementManagementSystem
{
    public partial class MaintainPaper : Form
    {
        public MaintainPaper()
        {
            InitializeComponent();
        }
        int paperId;
        private void MaintainPaper_Load(object sender, EventArgs e)
        {
            cmbRanking.Items.Add("1");
            cmbRanking.Items.Add("2");
            cmbRanking.Items.Add("3");
            cmbRanking.Items.Add("4");
            cmbRanking.Items.Add("5");
            cmbRanking.Items.Add("6及以后");
        }
        private void btnQuery_Click(object sender, EventArgs e)
        {
            //根据用户输入学生姓名查询该学生全部论文成果并展示查询结果
            string name = txtQueryName.Text;
            DataTable dt = PaperManager.GetALLPaper(name);
            dgvPaper.DataSource = dt;
        }
        private void dgvPaper_CellContentClick(object sender, DataGridViewCellEventArgs e)
        {
            //保存论文成果ID
            paperId = Convert.ToInt32(dgvPaper.CurrentRow.Cells["ID"].Value);
            //用户单击数据后为论文成果全部控件赋值
            txtNumber.Text = dgvPaper.CurrentRow.Cells["学号"].Value.ToString();
            txtName.Text = dgvPaper.CurrentRow.Cells["姓名"].Value.ToString();
            cmbRanking.Text = dgvPaper.CurrentRow.Cells["作者排名"].Value.ToString();
            txtJournalName.Text = dgvPaper.CurrentRow.Cells["期刊名称"].Value.ToString();
            txtPaperName.Text = dgvPaper.CurrentRow.Cells["论文名称"].Value.ToString();
            dtpTime.Text = dgvPaper.CurrentRow.Cells["发表时间"].Value.ToString();
        }
        private void btnUpdate_Click(object sender, EventArgs e)
        {
            //创建论文成果对象,并赋值
            Paper paper = new Paper();
```

```csharp
            paper.PaperId = paperId;
            paper.StuNum = Convert.ToInt32(txtNumber.Text);
            paper.StuName = txtName.Text;
            paper.StuRanking = Convert.ToInt32(cmbRanking.Text);
            paper.JournalName = txtJournalName.Text;
            paper.PaperName = txtPaperName.Text;
            paper.PaperTime = dtpTime.Text;
            //修改论文成果信息
            if (PaperManager.UpdatePaper(paper))
            {
                MessageBox.Show("论文成果信息修改成功！");
                //更新展示查询结果
                DataTable dt = PaperManager.GetALLPaper(txtQueryName.Text);
                dgvPaper.DataSource = dt;
            }
            else
            {
                MessageBox.Show("论文成果信息修改失败！");
            }
        }
        private void btnDelete_Click(object sender, EventArgs e)
        {
            if (MessageBox.Show("确定要删除论文成果信息吗？", "提示", MessageBoxButtons.OKCancel, MessageBoxIcon.Information) == DialogResult.OK)
            {
                //删除论文成果信息
                if (PaperManager.DeletePaper(paperId))
                {
                    MessageBox.Show("论文成果信息删除成功！");
                    //更新展示查询结果
                    DataTable dt = PaperManager.GetALLPaper(txtQueryName.Text);
                    dgvPaper.DataSource = dt;
                }
                else
                {
                    MessageBox.Show("论文成果信息删除失败！");
                }
            }
        }
    }
}
```

7.7.8 添加竞赛成果设计

1. 窗体设计

"添加竞赛成果"窗体如图7-9所示。

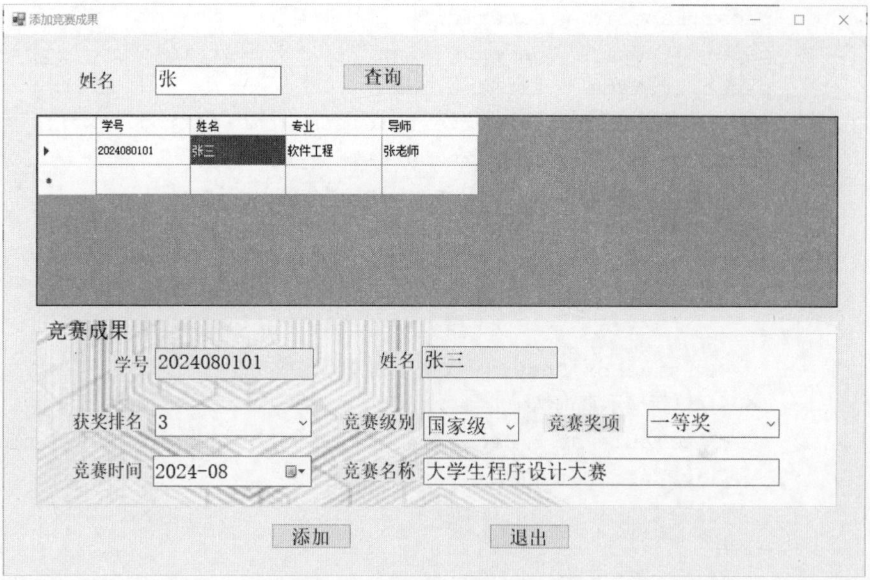

图 7-9 "添加竞赛成果"窗体

"添加竞赛成果"窗体和主要控件的属性设置如表 7-13 所示。

表 7-13 "添加竞赛成果"窗体和主要控件的属性设置

Name	类 型	Text 属性	其 他 属 性
AddCompetition	Form	添加竞赛成果	
txtQueryName	TextBox		
btnQuery	Button	查询	
dgvStudent	DataGridView		
groupBox1	GroupBox	竞赛成果	
txtNumber	TextBox		ReadOnly：True
txtName	TextBox		ReadOnly：True
cmbRanking	ComboBox		
cmbLevel	ComboBox		
cmbPrize	ComboBox		
dtpTime	DateTimePicker		
txtComName	TextBox		
btnAdd	Button	添加	
btnExit	Button	退出	

2. 功能实现代码

```
using System;
using System.Data;
using System.Windows.Forms;
using BLL;
using Models;
```

```csharp
namespace StudentAchievementManagementSystem
{
    public partial class AddCompetition : Form
    {
        public AddCompetition()
        {
            InitializeComponent();
        }
        private void btnQuery_Click(object sender, EventArgs e)
        {
            //根据用户输入学生姓名模糊查询并展示查询结果
            string name = txtQueryName.Text;
            DataTable dt = StudentManager.GetALLStudent(name);
            dgvStudent.DataSource = dt;
        }
        private void AddCompetition_Load(object sender, EventArgs e)
        {
            //获奖排名下拉列表框加载值
            cmbRanking.Items.Add("1");
            cmbRanking.Items.Add("2");
            cmbRanking.Items.Add("3");
            cmbRanking.Items.Add("4");
            cmbRanking.Items.Add("5");
            cmbRanking.Items.Add("6及以后");
            //竞赛级别下拉列表框加载值
            cmbLevel.Items.Add("国际级");
            cmbLevel.Items.Add("国家级");
            cmbLevel.Items.Add("省级");
            cmbLevel.Items.Add("校级");
            //竞赛奖项下拉列表框加载值
            cmbPrize.Items.Add("特等奖");
            cmbPrize.Items.Add("一等奖");
            cmbPrize.Items.Add("二等奖");
            cmbPrize.Items.Add("三等奖");
        }
        private void dgvStudent_CellContentClick(object sender, DataGridViewCellEventArgs e)
        {
            //用户单击数据后为学号和姓名文本框赋值
            txtNumber.Text = dgvStudent.CurrentRow.Cells["学号"].Value.ToString();
            txtName.Text = dgvStudent.CurrentRow.Cells["姓名"].Value.ToString();
        }
        private void btnExit_Click(object sender, EventArgs e)
        {
            this.Close();
```

```
        }
        private void btnAdd_Click(object sender, EventArgs e)
        {
            // 1.创建竞赛成果对象,并赋值
            Competition competition = new Competition();
            competition.StuNum = Convert.ToInt32(txtNumber.Text);
            competition.StuName = txtName.Text;
            competition.StuRanking = Convert.ToInt32(cmbRanking.Text);
            competition .ComName = txtComName.Text;
            competition.ComLevel = cmbLevel.Text;
            competition .ComPrize = cmbPrize.Text;
            competition.ComTime = dtpTime.Text;
            // 2.存入数据库表
            if (CompetitionManager.AddCompetition(competition))
            {
                MessageBox.Show("添加竞赛成果成功!");
            }
            else
            {
                MessageBox.Show("添加竞赛成果失败!");
            }
        }
    }
}
```

7.7.9 维护竞赛成果设计

1. 窗体设计

"维护竞赛成果"窗体如图 7-10 所示。

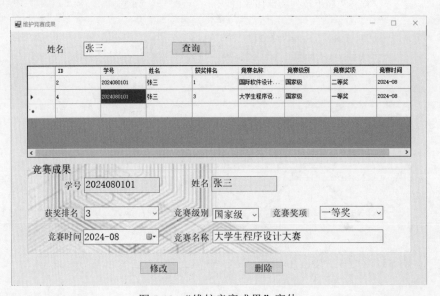

图 7-10 "维护竞赛成果"窗体

"维护竞赛成果"窗体和主要控件的属性设置如表 7-14 所示。

表 7-14 "维护竞赛成果"窗体和主要控件的属性设置

Name	类 型	Text 属性	其 他 属 性
MaintainCompetition	Form	维护竞赛成果	
txtQueryName	TextBox		
btnQuery	Button	查询	
dgvCompetition	DataGridView		
groupBox1	GroupBox	竞赛成果	
txtNumber	TextBox		ReadOnly：True
txtName	TextBox		ReadOnly：True
cmbRanking	ComboBox		
cmbLevel	ComboBox		
cmbPrize	ComboBox		
dtpTime	DateTimePicker		
txtCompetition	TextBox		
btnUpdate	Button	修改	
btnDelete	Button	删除	

2. 功能实现代码

```
using System;
using System.Data;
using System.Windows.Forms;
using BLL;
using Models;

namespace StudentAchievementManagementSystem
{
    public partial class MaintainCompetition : Form
    {
        public MaintainCompetition()
        {
            InitializeComponent();
        }
        int comId;
        private void MaintainCompetition_Load(object sender, EventArgs e)
        {
            //获奖排名下拉列表框加载值
            cmbRanking.Items.Add("1");
            cmbRanking.Items.Add("2");
            cmbRanking.Items.Add("3");
            cmbRanking.Items.Add("4");
            cmbRanking.Items.Add("5");
```

```csharp
            cmbRanking.Items.Add("6及以后");
            //"竞赛级别"下拉列表框加载值
            cmbLevel.Items.Add("国际级");
            cmbLevel.Items.Add("国家级");
            cmbLevel.Items.Add("省级");
            cmbLevel.Items.Add("校级");
            //"竞赛奖项"下拉列表框加载值
            cmbPrize.Items.Add("特等奖");
            cmbPrize.Items.Add("一等奖");
            cmbPrize.Items.Add("二等奖");
            cmbPrize.Items.Add("三等奖");
        }
        private void btnQuery_Click(object sender, EventArgs e)
        {
            //根据用户输入学生姓名查询该学生全部竞赛成果并展示查询结果
            string name = txtQueryName.Text;
            DataTable dt = CompetitionManager.GetALLCompetition(name);
            dgvCompetition.DataSource = dt;
        }
        private void dgvCompetition_CellContentClick(object sender, DataGridViewCellEventArgs e)
        {
            //保存竞赛成果ID
            comId = Convert.ToInt32(dgvCompetition.CurrentRow.Cells["ID"].Value);
            //用户单击数据后为竞赛成果全部控件赋值
            txtNumber.Text = dgvCompetition.CurrentRow.Cells["学号"].Value.ToString();
            txtName.Text = dgvCompetition.CurrentRow.Cells["姓名"].Value.ToString();
            cmbRanking.Text = dgvCompetition.CurrentRow.Cells["获奖排名"].Value.ToString();
            txtComName.Text = dgvCompetition.CurrentRow.Cells["竞赛名称"].Value.ToString();
            cmbLevel.Text = dgvCompetition.CurrentRow.Cells["竞赛级别"].Value.ToString();
            cmbPrize.Text = dgvCompetition.CurrentRow.Cells["竞赛奖项"].Value.ToString();
            dtpTime.Text = dgvCompetition.CurrentRow.Cells["竞赛时间"].Value.ToString();
        }
        private void btnUpdate_Click(object sender, EventArgs e)
        {
            //创建竞赛成果对象,并赋值
            Competition competition = new Competition();
```

```csharp
                competition.ComId = comId;
                competition.StuNum = Convert.ToInt32(txtNumber.Text);
                competition.StuName = txtName.Text;
                competition.StuRanking = Convert.ToInt32(cmbRanking.Text);
                competition.ComName = txtComName.Text;
                competition.ComLevel = cmbLevel.Text;
                competition.ComPrize = cmbPrize.Text;
                competition.ComTime = dtpTime.Text;
                //修改竞赛成果信息
                if (CompetitionManager.UpdateCompetition(competition))
                {
                    MessageBox.Show("竞赛成果信息修改成功！");
                    //更新展示查询结果
                    DataTable dt = CompetitionManager
.GetALLCompetition(txtQueryName.Text);
                    dgvCompetition.DataSource = dt;
                }
                else
                {
                    MessageBox.Show("竞赛成果信息修改失败！");
                }
        }
        private void btnDelete_Click(object sender, EventArgs e)
        {
            if (MessageBox.Show("确定要删除竞赛成果信息吗？", "提示",
MessageBoxButtons.OKCancel, MessageBoxIcon.Information) == DialogResult.OK)
            {
                //删除竞赛成果信息
                if (CompetitionManager.DeleteCompetition(comId))
                {
                    MessageBox.Show("竞赛成果信息删除成功！");
                    //更新展示查询结果
                    DataTable dt = CompetitionManager
.GetALLCompetition(txtQueryName.Text);
                    dgvCompetition.DataSource = dt;
                }
                else
                {
                    MessageBox.Show("竞赛成果信息删除失败！");
                }
            }
        }
    }
```

7.7.10 查询导师成果设计

1. 窗体设计

"查询导师成果"窗体如图 7-11 所示。

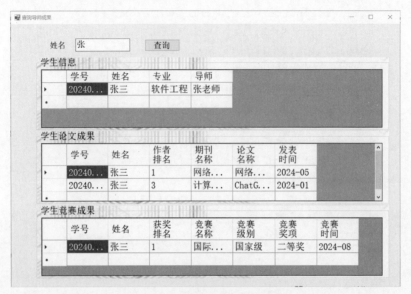

图 7-11 "查询导师成果"窗体

"查询导师成果"窗体和主要控件的属性设置如表 7-15 所示。

表 7-15 "查询导师成果"窗体和主要控件的属性设置

Name	类型	Text 属性	其他属性
TeacherSearch	Form	查询导师成果	
txtQueryName	TextBox		
btnQuery	Button	查询	
dgvStudent	DataGridView		
dgvPaper	DataGridView		
dgvCompetition	DataGridView		

2. 功能实现代码

```
using System;
using System.Data;
using System.Windows.Forms;
using BLL;

namespace StudentAchievementManagementSystem
{
    public partial class TeacherSearch : Form
    {
        public TeacherSearch()
```

```csharp
        {
            InitializeComponent();
        }
        private void btnQuery_Click(object sender, EventArgs e)
        {
            //根据用户输入导师姓名模糊查询学生信息并展示查询结果
            string name = txtQueryName.Text;
            DataTable dt = StudentManager.GetALLStudentByTeaName(name);
            dgvStudent.DataSource = dt;
        }
        private void dgvStudent_CellContentClick(object sender, DataGridViewCellEventArgs e)
        {
            //存储学生学号
            int stuNum = Convert.ToInt32(dgvStudent.CurrentRow.Cells["学号"].Value);
            //根据学号查询该学生全部论文成果信息并展示
            dgvPaper.DataSource = PaperManager.GetALLPaperByNum(stuNum);
            //根据学号查询该学生全部竞赛成果信息并展示
            dgvCompetition.DataSource = CompetitionManager.GetALLCompetitionByNum(stuNum);
        }
    }
}
```

7.7.11 查询学生成果设计

1. 窗体设计

"查询学生成果"窗体如图 7-12 所示。

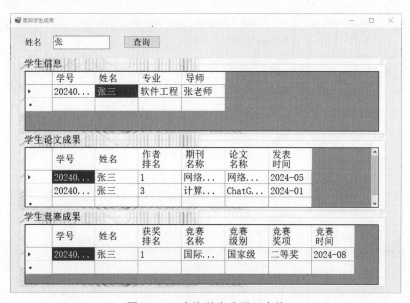

图 7-12 "查询学生成果"窗体

"查询学生成果"窗体和主要控件的属性设置如表 7-16 所示。

表 7-16 "查询学生成果"窗体和主要控件的属性设置

Name	类 型	Text 属性	其他属性
StudentSearch	Form	查询学生成果	
txtQueryName	TextBox		
btnQuery	Button	查询	
dgvStudent	DataGridView		
dgvPaper	DataGridView		
dgvCompetition	DataGridView		

2. 功能实现代码

```
using System;
using System.Data;
using System.Windows.Forms;
using BLL;
namespace StudentAchievementManagementSystem
{
    public partial class StudentSearch : Form
    {
        public StudentSearch()
        {
            InitializeComponent();
        }
        private void btnQuery_Click(object sender, EventArgs e)
        {
            //根据用户输入学生姓名模糊查询并展示查询结果
            string name = txtQueryName.Text;
            DataTable dt = StudentManager.GetALLStudent(name);
            dgvStudent.DataSource = dt;
        }
        private void dgvStudent_CellContentClick(object sender, DataGridViewCellEventArgs e)
        {
            //存储学生学号
            int stuNum = Convert.ToInt32(dgvStudent.CurrentRow.Cells["学号"].Value);
            //根据学号查询该学生全部论文成果信息并展示
            dgvPaper.DataSource = PaperManager.GetALLPaperByNum(stuNum);
            //根据学号查询该学生全部竞赛成果信息并展示
            dgvCompetition.DataSource = CompetitionManager.GetALLCompetitionByNum(stuNum);
        }
    }
}
```

第8章　实践项目——酒店管理系统

本章作为课程设计的中级案例，使用 ASP.NET 与 MySQL 设计一个简单的酒店管理系统。酒店管理系统是一种可以提高酒店管理效率的软件，一般包含前台接待、客房管理、住客管理等功能。

8.1 需求分析

酒店管理系统主要功能如下：

（1）登录。用户在登录时，需要输入用户名和密码，登录成功后，根据用户的权限显示相应的主界面。系统中用户权限划分成管理员用户和员工用户，管理员用户可以进行用户管理、客房管理、查询管理等操作；员工用户只可以进行客房管理与查询管理等操作。

（2）用户管理。用户管理功能可以进行用户的添加、修改和删除操作。添加新用户时，需要输入用户的用户名、密码，并选择用户角色。

（3）客房管理。客房管理功能可以进行入住登记和退房登记。住客入住时，工作人员需要选择空房间，并记录入住日期、离开日期、住客姓名、性别、证件类型、证件号码、证件地址、联系电话、预收押金以及住宿人数等信息。住客退房时，工作人员需要选择房号，并进行费用结算，添加备注信息等。

（4）查询管理。查询管理功能可以进行住客信息查询和客房信息查询。住客信息可以根据姓名进行查询，并可以对住客姓名、性别、联系电话、证件类型、证件号码以及证件地址等信息进行修改。客户信息查询可根据房间信息查询出该房间的住客信息，包括姓名、入住时间、退房时间等。

8.2 概要设计

8.2.1 架构设计

酒店管理系统采用三层架构设计，分为数据访问层（data access layer，DAL）、业务逻辑层（business logic layer，BLL）以及表示层（user interface layer，UIL）。数据访问层用于数据存储与数据访问操作；业务逻辑层包含与核心业务相关的逻辑流程，主要实现业务规则与业务逻辑；表示层主要完成与用户交互任务，并将相关数据提交给业务逻辑层来处理。

8.2.2　功能设计

根据需求分析，酒店管理系统包括管理员用户和员工用户两类。管理员用户可以进行用户管理、客房管理、查询管理等操作，员工用户可以进行客房管理与查询管理等操作。

酒店管理系统功能模块设计如图 8-1 所示。

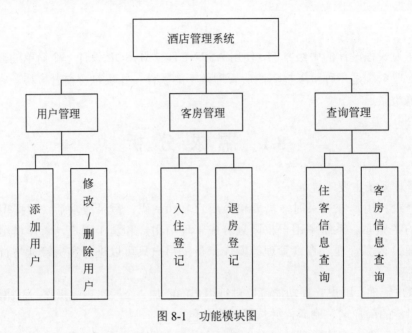

图 8-1　功能模块图

8.3　数据库设计

酒店管理系统采用 MySQL 数据库，名称为 hotel，其中共 4 个数据表，分别是用户信息表（user）、房间信息表（room）、入住信息表（checkin）、退房信息表（checkout）。

1．用户信息表

用户信息表用来存储管理员和员工的相关信息，包括用户名、用户密码及权限等字段，数据表的结构如表 8-1 所示。

表 8-1　用户信息表（user）

字 段 名	数 据 类 型	允 许 空 值	字 段 说 明
userName	varchar（20）	否	用户名（主键）
userPwd	varchar（20）	否	用户密码
role	int（2）	否	权限

2．房间信息表

房间信息表用来存储酒店房间的相关信息，包括房间号、房间类型、楼层、价格、可入住人数、已入住人数及备注等字段，数据表的结构如表 8-2 所示。

表 8-2 房间信息表（room）

字 段 名	数 据 类 型	允 许 空 值	字 段 说 明
roomId	varchar（30）	否	房间号（主键）
roomType	varchar（30）	否	房间类型
roomFloor	varchar（20）	否	楼层
price	varchar（10）	否	价格
personNum	int（11）	否	可入住人数
inPerson	int（11）	否	已入住人数
note	varchar（250）	是	备注

3．入住信息表

入住信息表主要存储住客的相关信息，包括入住编号、房间号、价格、押金、入住时间、离开时间、住客姓名、性别、电话、证件类型、证件号码、证件住址、入住人数、操作员及是否退房等字段，数据表的结构如表 8-3 所示。

表 8-3 入住信息表（checkin）

字 段 名	数 据 类 型	允 许 空 值	字 段 说 明
inId	int（11）	否	入住编号（主键，自动递增）
roomId	varchar（30）	否	房间号
price	varchar（5）	否	价格
foregift	varchar（5）	否	押金
inTime	datetime	否	入住时间
outTime	datetime	否	离开时间
clientName	varchar（30）	否	住客姓名
gender	bit（1）	否	性别
phone	varchar（30）	否	电话
certType	varchar（20）	否	证件类型
certId	varchar（30）	否	证件号码
address	varchar（50）	否	证件住址
personNum	int（11）	否	入住人数
oper	varchar（20）	否	操作员
delMark	bit（1）	否	是否退房

4．退房信息表

退房信息表用来存储退房的相关信息，包括退房编号、入住编号、退房时间、房间号、住客姓名、入住时间、价格、押金、费用总额、结账金额、备注及操作员等字段，数据表的结构如表 8-4 所示。

表 8-4　退房信息表（checkout）

字 段 名	数 据 类 型	允 许 空 值	字 段 说 明
outId	varchar（14）	否	退房编号（主键）
inId	int（11）	否	入住编号
outTime	datetime	否	退房时间
roomId	varchar（30）	否	房间号
clientName	varchar（30）	否	住客姓名
inTime	datetime	否	入住时间
price	varchar（5）	否	价格
foregift	varchar（5）	否	押金
total	varchar（10）	否	费用总额
account	varchar（10）	否	结账金额
note	varchar（250）	是	备注
oper	varchar（20）	否	操作员

8.4　实体模型设计

实体模型包含了所有的数据信息，这些数据信息以各种 Entity 实例的形式存在。实体模型类库 Models 由 4 个核心类组成，分别为 User 类、RoomInfo 类、CheckInRoom 类和 CheckOutRoom 类。

1．User 类

User 类用于封装 hotel 数据库中的 user 数据表的用户数据，程序代码如下：

```csharp
public class User
    {
        private string _name;
        private string _pwd;
        private int _role;
        public string Name
        {
            get { return _name; }
            set { _name = value; }
        }
        public string Pwd
        {
            get { return _pwd; }
            set { _pwd = value; }
        }
        public int Role
        {
            get { return _role; }
            set { _role = value; }
        }
```

```
    public User()
    {
        _name = "";
        _pwd = "";
        _role = 0;
    }
    public User(string name, string pwd, int role)
    {
        _name = name;
        _pwd = pwd;
        _role = role;
    }
}
```

2. RoomInfo 类

RoomInfo 类用于封装 hotel 数据库中的 room 数据表的房间数据，程序代码如下：

```
public class RoomInfo
    {
    private string _roomId;
    private string _roomType;
    private string _roomFloor;
    private double _price;
    private int _personNum;
    private int _inPerson;
    private string _note;
    public string RoomId
    {
        get { return _roomId; }
        set { _roomId = value; }
    }
    public string RoomType
    {
        get { return _roomType; }
        set { _roomType = value; }
    }
    public string RoomFloor
    {
        get { return _roomFloor; }
        set { _roomFloor = value; }
    }
    public double Price
    {
        get { return _price; }
        set { _price = value; }
    }
    public int PersonNum
    {
        get { return _personNum; }
```

```
            set { _personNum = value; }
        }
        public int InPerson
        {
            get { return _inPerson; }
            set { _inPerson = value; }
        }
        public string Note
        {
            get { return _note; }
            set { _note = value; }
        }
    }
```

3. CheckInRoom 类

CheckInRoom 类用于封装 hotel 数据库中的 checkin 数据表的入住登记数据,程序代码如下:

```
    public class CheckInRoom
        {
            private int _inId;
            private string _roomId;
            private double _price;
            private double _foregift;
            private DateTime _inTime;
            private DateTime _outTime;
            private string _clientName;
            private int _gender;
            private string _phone;
            private string _certType;
            private string _certId;
            private string _address;
            private int _personNum;
            private string _oper;
            private int _delMark;
            public int InId
            {
                get { return _inId; }
                set { _inId = value; }
            }
            public string RoomId
            {
                get { return _roomId; }
                set { _roomId = value; }
            }
            public double Price
            {
                get { return _price; }
```

```csharp
        set { _price = value; }
    }
    public double Foregift
    {
        get { return _foregift; }
        set { _foregift = value; }
    }
    public DateTime InTime
    {
        get { return _inTime; }
        set { _inTime = value; }
    }
    public DateTime OutTime
    {
        get { return _outTime; }
        set { _outTime = value; }
    }
    public string ClientName
    {
        get { return _clientName; }
        set { _clientName = value; }
    }
    public int Gender
    {
        get { return _gender; }
        set { _gender = value; }
    }
    public string Phone
    {
        get { return _phone; }
        set { _phone = value; }
    }
    public string CertType
    {
        get { return _certType; }
        set { _certType = value; }
    }
    public string CertId
    {
        get { return _certId; }
        set { _certId = value; }
    }
    public string Address
    {
        get { return _address; }
        set { _address = value; }
    }
    public int PersonNum
```

```csharp
        {
            get { return _personNum; }
            set { _personNum = value; }
        }
        public string Oper
        {
            get { return _oper; }
            set { _oper = value; }
        }
        public int DelMark
        {
            get { return _delMark; }
            set { _delMark = value; }
        }
    }
```

4. CheckOutRoom 类

CheckOutRoom 类用于封装 hotel 数据库中的 checkout 数据表的退房登记数据，程序代码如下：

```csharp
    public class CheckOutRoom
    {
        private string _outId;
        private int _inId;
        private DateTime _outTime;
        private string _roomId;
        private string _clientName;
        private DateTime _inTime;
        private double _price;
        private double _foregift;
        private double _total;
        private double _account;
        private string _note;
        private string _oper;
        public string OutId
        {
            get { return _outId; }
            set { _outId = value; }
        }
        public int InId
        {
            get { return _inId; }
            set { _inId = value; }
        }
        public DateTime OutTime
        {
            get { return _outTime; }
            set { _outTime = value; }
```

```csharp
    }
    public string RoomId
    {
        get { return _roomId; }
        set { _roomId = value; }
    }
    public string ClientName
    {
        get { return _clientName; }
        set { _clientName = value; }
    }
    public DateTime InTime
    {
        get { return _inTime; }
        set { _inTime = value; }
    }
    public double Price
    {
        get { return _price; }
        set { _price = value; }
    }
    public double Foregift
    {
        get { return _foregift; }
        set { _foregift = value; }
    }
    public double Total
    {
        get { return _total; }
        set { _total = value; }
    }
    public double Account
    {
        get { return _account; }
        set { _account = value; }
    }
    public string Note
    {
        get { return _note; }
        set { _note = value; }
    }
    public string Oper
    {
        get { return _oper; }
        set { _oper = value; }
    }
}
```

8.5 数据访问层设计

数据访问层负责加载数据库驱动程序、创建和关闭数据库连接以及对数据库表进行添加、修改、删除和查询等操作。数据访问类库 DAL 由一个核心类 DBOper.cs 组成，其中封装了执行数据的增、删、改、查等功能的方法。

1. 配置 App.config

为了方便数据操作，将一些配置参数存储在窗体项目 App.config 配置文件中。参考程序代码如下：

```xml
<configuration>
    <connectionStrings>
        <add name = "HotelConStr" connectionString = "server = localhost;user id = root;password = 123456;database = hotel;Charset = utf8;" />
    </connectionStrings>
</configuration>
```

2. DBOper 类中的 Connection 属性

在 DBOper 类中声明了一个连接对象只读属性，当读取属性值时，将自动连接数据库，程序代码如下：

```csharp
private static MySqlConnection conn;
public static MySqlConnection Connection
{
    get
    {
        if (conn == null)
        {
            string connStr = ConfigurationManager.ConnectionStrings["HotelConStr"].ConnectionString;
            conn = new MySqlConnection(connStr);
            conn.Open();
        }
        else if (conn.State == ConnectionState.Closed)
        {
            conn.Open();
        }
        else if (conn.State == ConnectionState.Broken)
        {
            conn.Close();
            conn.Open();
        }
        return conn;
    }
}
```

3. DBOper 类中的 ExecuteCommand()方法

ExecuteCommand() 方法是自定义的方法重载，ExecuteCommand() 方法中的 MySqlCommand 类表示要对 MySQL 数据库执行一个 SQL 语句，返回值为 int 型，表示受影响的行数，主要实现对数据库中的数据进行不带参数和带参数的添加、修改、删除等功能。

```
public static int ExecuteCommand(string sql)
{
    MySqlCommand cmd = new MySqlCommand(sql, Connection);
    return cmd.ExecuteNonQuery();
}
public static int ExecuteCommand(string sql, params MySqlParameter[] values)
{
    MySqlCommand cmd = new MySqlCommand(sql, Connection);
    cmd.Parameters.AddRange(values);
    return cmd.ExecuteNonQuery();
}
```

4. DBOper 类中的 GetReader()方法

GetReader()方法是自定义的方法重载，GetReader()方法返回 MySqlDataReader 类型的数据，主要实现对数据库中的数据进行不带参数和带参数的读取功能。

```
public static MySqlDataReader GetReader(string sql)
{
    MySqlCommand cmd = new MySqlCommand(sql, Connection);
    return cmd.ExecuteReader();
}
public static MySqlDataReader GetReader(string sql, params MySqlParameter[] values)
{
    MySqlCommand cmd = new MySqlCommand(sql, Connection);
    cmd.Parameters.AddRange(values);
    return cmd.ExecuteReader();
}
```

5. DBOper 类中的 GetDataTable()方法

GetDataTable()方法是自定义的方法重载，GetDataTable()方法返回 DataTable 类型的数据，主要实现对数据库中的数据表进行不带参数和带参数的读取功能。

```
public static DataTable GetDataTable(string sql)
{
    MySqlDataAdapter da = new MySqlDataAdapter(sql, Connection);
    DataSet ds = new DataSet();
    da.Fill(ds);
    return ds.Tables[0];
}
public static DataTable GetDataTable(string sql, params MySqlParameter[] values)
```

```csharp
{
    MySqlCommand cmd = new MySqlCommand(sql, Connection);
    cmd.Parameters.AddRange(values);
    MySqlDataAdapter da = new MySqlDataAdapter(cmd);
    DataSet ds = new DataSet();
    da.Fill(ds);
    return ds.Tables[0];
}
```

8.6 业务逻辑层设计

业务逻辑层类库 BLL 由 4 个核心类组成，分别为 RoleManager.cs、UserManager.cs、RoomManager.cs 和 ClientManager.cs。

1. RoleManager 类

RoleManager 类，用于保存当前登录用户的相关信息。

```csharp
public class RoleManager
{
    public static User curUser = new User();
}
```

【分析】

用户登录成功后，把从 user（用户信息表）中查询到的数据（User 类型对象）存储到 RoleManager 类的 curUser 字段中，用于在入住登记等界面中获取当前用户（即操作员）。

2. UserManager 类

UserManager 类用于对用户进行管理。该类封装了 5 个方法，分别为 GetUser()方法、AddUser()方法、GetUserInfo()方法、UpdateUser()方法和 DeleteUser()方法。

（1）GetUser()方法。用于根据用户名和密码获取用户，代码如下：

```csharp
public static User GetUser(string name,string pwd)
{
    User user = null;
    string sql = "select * from user where userName=\'" + name + "\' and userPwd = \'" + pwd + "\'";
    MySqlDataReader reader = DBOper.GetReader(sql);
    if (reader.Read())
    {
        user = new User();
        user.Name = Convert.ToString(reader["userName"]);
        user.Pwd = Convert.ToString(reader["userPwd"]);
        user.Role = Convert.ToInt32(reader["role"]);
    }
    reader.Close();
    return user;
}
```

【分析】

用户在登录时，输入用户名和密码，用户名和密码作为 GetUser()方法参数，然后从数据库 user（用户信息表）中查询该用户，如果存在，则返回 User 类型对象。

（2）AddUser()方法。用于将新用户信息添加到数据库的 user 表中，代码如下：

```
public static bool AddUser(User user)
{
    string sql = "insert into user values(@name,@pwd,@role)";
    MySqlParameter p1 = new MySqlParameter("@name",user.Name);
    MySqlParameter p2 = new MySqlParameter("@pwd", user.Pwd);
    MySqlParameter p3 = new MySqlParameter("@role", user.Role);
    if (DBOper.ExecuteCommand(sql, p1, p2, p3) == 1)
    {
        return true;
    }
    else
    {
        return false;
    }
}
```

【分析】

在管理员添加新用户时，输入的用户名、密码和权限数据存储在 User 类型对象属性中，并作为 AddUser()方法的参数传入，如果添加成功，则返回 true，否则返回 false。

（3）GetUserInfo()方法。用于从 user 表中获取用户信息，返回值为 DataTable，代码如下：

```
public static DataTable GetUserInfo()
{
    string sql = @"select userName as 用户名,userPwd as 用户密码,role as 用户权限 from user";
    DataTable dt = DBOper.GetDataTable(sql);
    return dt;
}
```

【分析】

在管理员修改或删除用户时，在界面加载时将所有用户信息获取并返回到数据表中。

（4）UpdateUser()方法。用于保存修改后的用户信息。

```
public static bool UpdateUser(User user)
{
    string sqlUpdateUser = @"update user set userPwd = @userPwd,role = @role where userName = @userName";
    MySqlParameter p1 = new MySqlParameter("@userPwd", user.Pwd);
    MySqlParameter p2 = new MySqlParameter("@role", user.Role);
    MySqlParameter p3 = new MySqlParameter("@userName", user.Name);
    MySqlParameter[] paramArray = new MySqlParameter[] { p1, p2, p3 };
    if (DBOper.ExecuteCommand(sqlUpdateUser, paramArray) == 1)
    {
        return true;
```

```
            }
            else
            {
                return false;
            }
        }
```

【分析】

在管理员修改或删除用户时,单击"修改"按钮,将修改后的数据封装成 User 类型的对象,并执行数据库修改操作,其中用户名为主键,不能修改。

(5) DeleteUser()方法。用于删除用户信息。

```
public static bool DeleteUser(User user)
{
    string sqlDeleteUser = @"delete from user where userName = @userName";
    MySqlParameter p1 = new MySqlParameter("@userName", user.Name);
    MySqlParameter[] paramArray = new MySqlParameter[] { p1 };
    if (DBOper.ExecuteCommand(sqlDeleteUser, paramArray) == 1)
    {
        return true;
    }
    else
    {
        return false;
    }
}
```

【分析】

在管理员修改或删除用户时,单击"删除"按钮,将查询到的用户删除。

3. RoomManager 类

RoomManager 类用于对房间进行管理。该类中封装了 7 个方法,分别为 GetRoomInfo()方法、InsertRoomInfo()方法、GetUseRoomInfo()方法、GetInId()方法、InsertCheckOutRoomInfo()方法、GetRoomInfo()方法和 GetRoomUseInfo()方法。

(1) GetRoomInfo()方法。用于从 room 表中获取空房信息,返回值为 DataTable 类型,程序代码如下:

```
public static DataTable GetRoomInfo()
{
    string sql = @"select roomId as 房间号,roomType as 房间类型, roomFloor as 层数, price as 价格,personNum as 可入住人数,inPerson as 已入住人数,note as 备注 from room where inPerson = 0";
    DataTable dt = DBOper.GetDataTable(sql);
    return dt;
}
```

【分析】

在入住登记时,入住登记界面中将显示全部房间信息,即从 room(房间信息表)中查询入住人数 inPerson 值为 0 的数据,并将查询到的数据表返回。

（2）InsertRoomInfo()方法。用于保存入住信息到 checkin 表中，并且更新 room 表中的入住人数 inPerson 值，返回值为 bool 类型，程序代码如下：

```csharp
public static bool InsertRoomInfo(CheckInRoom checkInRoomInfo)
{
    string sqlInsert = @"insert into checkin values(null,@roomId,@price,
    @foregift,@inTime,@outTime,@clientName,@gender,@phone,@certType,@certId,
    @address,@personNum,@Oper,@delMark)";
    MySqlParameter p1 = new MySqlParameter("@roomId", checkInRoomInfo
    .RoomId);
    MySqlParameter p2 = new MySqlParameter("@price", checkInRoomInfo.Price);
    MySqlParameter p3 = new MySqlParameter("@foregift", checkInRoomInfo
    .Foregift);
    MySqlParameter p4 = new MySqlParameter("@inTime", checkInRoomInfo.InTime);
    MySqlParameter p5 = new MySqlParameter("@outTime", checkInRoomInfo
    .OutTime);
    MySqlParameter p6 = new MySqlParameter("@clientName", checkInRoomInfo
    .ClientName);
    MySqlParameter p7 = new MySqlParameter("@gender", checkInRoomInfo.Gender);
    MySqlParameter p8 = new MySqlParameter("@phone", checkInRoomInfo.Phone);
    MySqlParameter p9 = new MySqlParameter("@certType", checkInRoomInfo
    .CertType);
    MySqlParameter p10 = new MySqlParameter("@certId", checkInRoomInfo
    .CertId);
    MySqlParameter p11 = new MySqlParameter("@address", checkInRoomInfo
    .Address);
    MySqlParameter p12 = new MySqlParameter("@personNum", checkInRoomInfo
    .PersonNum);
    MySqlParameter p13 = new MySqlParameter("@Oper", checkInRoomInfo.Oper);
    MySqlParameter p14 = new MySqlParameter("@delMark", checkInRoomInfo
    .DelMark);
    MySqlParameter[] paramArray = new MySqlParameter[] { p1, p2, p3, p4, p5,
    p6, p7, p8, p9, p10, p11, p12, p13, p14 };
    string sqlUpdate = "update room set inPerson = @inPerson where roomId =
    @roomId";
    MySqlParameter n1 = new MySqlParameter("@roomId", checkInRoomInfo.RoomId);
    MySqlParameter n2 = new MySqlParameter("@inPerson", checkInRoomInfo
    .PersonNum);
    if (DBOper.ExecuteCommand(sqlInsert, paramArray) == 1 && DBOper
    .ExecuteCommand(sqlUpdate, n1, n2) == 1)
    {
        return true;
    }
    else
    {
        return false;
    }
}
```

【分析】

在入住登记时，操作员在入住登记界面中输入住客的相关信息，将信息添加到 checkin（入住信息表）中，并且将入住人数保存到 room（房间信息表）中的入住人数 inPerson 字段中，即表明该房间已不是空房间。如果以上两个表的操作都成功，则返回 true，否则返回 false。

（3）GetUseRoomInfo()方法。用于从 checkin 表中获取入住信息，其中 delMark=0 表示已入住并且没有退房，返回值为 DataTable 类型，程序代码如下：

```
public static DataTable GetUseRoomInfo()
{
    string sql = @"select roomId as 房间号,price as 价格,foregift as 押金,
    inTime as 入住时间,outTime as 退房时间,clientName as 住客姓名, gender as 性
    别,phone as 电话号码,certType as 证件类型, certId as 证件号码, address as 地
    址,personNum as 入住人数, oper as 登记者 from checkin where delMark=0";
    DataTable dt = DBOper.GetDataTable(sql);
    return dt;
}
```

【分析】

在退房登记时，在退房登记界面初始化时，需要获取所有已入住并且没有退房的入住信息，即从 checkin（入住信息表）中查询 delMark 值为 0 的全部数据并返回。

（4）GetInId()方法。用于从 checkin 表中根据房间号 roomId 查找 delMark=0 的退房房间的 inId，返回值为 DataTable 类型，程序代码如下：

```
public static DataTable GetInID(string roomId)
{
    string sql = @"select inId from checkin where delMark=0 and roomId=" + roomId;
    DataTable dt = DBOper.GetDataTable(sql);
    return dt;
}
```

【分析】

在退房登记时，当操作员单击"保存"按钮时，要将退房信息保存，其中 inId 的值要在 checkin（入住信息表）中根据房间号 roomId（且 delMark 为 0，即当前已入住并且没有退房）字段查询并返回。

（5）InsertCheckOutRoomInfo()方法。用于保存退房信息到 checkout 表中，更新 checkin 表中的 delMark 值为 1，并且更新 room 表中的 InPerson 为 0，返回值为 bool 类型，程序代码如下：

```
public static bool InsertCheckOutRoomInfo(CheckOutRoom checkOutRoomInfo)
{
    string sqlInsert = @"insert into checkout values(@outId,@inId,@outTime,
    @roomId,@clientName,@inTime,@price,@foregift,@total,@account,@note,
    @oper)";
```

```csharp
string sqlUpdateRegister = "update checkin set delMark = 1 where inId = @inId";
string sqlUpdateRoom = "update room set inPerson=0 where roomId = @roomId";
MySqlParameter p1 = new MySqlParameter("@outId", checkOutRoomInfo.OutId);
MySqlParameter p2 = new MySqlParameter("@inId", checkOutRoomInfo.InId);
MySqlParameter p3 = new MySqlParameter("@roomId", checkOutRoomInfo.RoomId);
MySqlParameter p4 = new MySqlParameter("@price", checkOutRoomInfo.Price.ToString());
MySqlParameter p5 = new MySqlParameter("@foregift", checkOutRoomInfo.Foregift.ToString());
MySqlParameter p6 = new MySqlParameter("@total", checkOutRoomInfo.Total.ToString());
MySqlParameter p7 = new MySqlParameter("@account", checkOutRoomInfo.Account.ToString());
MySqlParameter p8 = new MySqlParameter("@inTime", checkOutRoomInfo.InTime);
MySqlParameter p9 = new MySqlParameter("@outTime", checkOutRoomInfo.OutTime);
MySqlParameter p10 = new MySqlParameter("@clientName", checkOutRoomInfo.ClientName);
MySqlParameter p11 = new MySqlParameter("@oper", checkOutRoomInfo.Oper);
MySqlParameter p12 = new MySqlParameter("@note", checkOutRoomInfo.Note);
MySqlParameter[] paramArray = new MySqlParameter[] { p1, p2, p3, p4, p5, p6, p7, p8, p9, p10, p11, p12 };
MySqlParameter n1 = new MySqlParameter("@inId", checkOutRoomInfo.InId);
MySqlParameter n2 = new MySqlParameter("@roomId", checkOutRoomInfo.RoomId);
if (DBOper.ExecuteCommand(sqlInsert, paramArray) == 1 &&
    DBOper.ExecuteCommand(sqlUpdateRegister, n1) == 1 &&
    DBOper.ExecuteCommand(sqlUpdateRoom, n2) == 1)
{
    return true;
}
else
{
    return false;
}
}
```

【分析】

在退房登记时，当操作员单击"保存"按钮时，将所有信息保存到 checkout（退房信息表）中，并且修改 checkin（入住信息表）中的 delMark 值为 1，即表示已退房，同时，修改 room（房间信息表）中的入住人数 inPerson 值为 0，即表示该房间为房间。如果对以上三个表的操作都成功，则返回 true，否则返回 false。

（6）GetRoomInfo()方法。用于从 room 表中根据房间号 roomId 查询房间信息，返回值为 DataTable 类型，程序代码如下：

```
public static DataTable GetRoomInfo(string roomId)
{
    string sql = @"select * from room where roomId = " + roomId;
    DataTable dt = DBOper.GetDataTable(sql);
    return dt;
}
```

【分析】

在住客信息查询界面中，操作员如果要修改查询到的住客信息的入住人数时，需要验证修改后的值不能大于该房间允许入住的人数，因此，需要在 room（房间信息表）中根据房间号 roomId 查询该房间信息并返回。

（7）GetRoomUseInfo()方法。用于从 room 表中查询非空房间信息，即 inPeron 值不为 0 的房间信息，返回值为 DataTable 类型，程序代码如下：

```
public static DataTable GetRoomUseInfo()
{
    string sql = @"select roomId as 房间号,roomType as 房间类型,roomFloor as 层数, price as 价格,personNum as 可入住人数,inPerson as 已入住人数, note as 备注 from room where inPerson != 0";
    DataTable dt = DBOper.GetDataTable(sql);
    return dt;
}
```

【分析】

在客户信息查询界面加载中，需要从 room（房间信息表）查询已入住（即入住人数 inPerson 值为 0）的房间信息。

4. ClientManager 类

ClientManager 类用于对住客信息进行管理。该类封装 4 个方法，分别为 GetClientInfo() 方法、GetAllClient()方法、UpdateRegister()方法和 GetClient()方法。

（1）GetClientInfo()方法。用于从 checkin 表中获取住客信息，返回值为 DataTable 类型，程序代码如下：

```
public static DataTable GetClientInfo()
{
    string sql = @"select roomId as 房间号,price as 价格,foregift as 押金, inTime as 入住时间,outTime as 退房时间,clientName as 住客姓名,gender as 性别,phone
```

```
        as 电话号码,certType as 证件类型,certId as 证件号码,address as 地址,personNum
        as 入住人数, oper as 登记者,delMark as 是否退房,inId from checkin";
    DataTable dt = DBOper.GetDataTable(sql);
    return dt;
}
```

【分析】

在住客信息查询界面加载时,要从 checkin(入住信息表)中查询所有数据。

(2)GetAllClient()方法。用于从 checkin 表中根据姓名模糊查询住客信息,返回值为 DataTable 类型,程序代码如下:

```
public static DataTable GetALLClient(string clientName)
{
    string sql = @"select roomId as 房间号,price as 价格,foregift as 押金, inTime
    as 入住时间,outTime as 退房时间,clientName as 住客姓名, gender as性别,phone
    as 电话号码,certType as 证件类型, certId as 证件号码,address as 地址,personNum as
    入住人数, oper as 登记者,delMark as 是否退房,inId from checkin where
    clientName like '%" + clientName + "%'";
    DataTable dt = DBOper.GetDataTable(sql);
    return dt;
}
```

【分析】

在住客信息查询界面中,可以根据操作员输入的住客姓名进行模糊查询。

(3)UpdateRegister()方法。用于在 checkin 表中保存住客信息,返回值为 bool 类型,程序代码如下:

```
public static bool UpdateRegister(CheckInRoom checkInRoomInfo)
{
    string sqlUpdateRegister = @"update checkin set clientName=@clientName,
    gender = @gender,phone = @phone,certType = @certType,certId = @certId,
    address = @address,personNum = @personNum where inId = @inId";
    MySqlParameter p1 = new MySqlParameter("@clientName", checkInRoomInfo
    .ClientName);
    MySqlParameter p2 = new MySqlParameter("@gender", checkInRoomInfo.Gender);
    MySqlParameter p3 = new MySqlParameter("@phone", checkInRoomInfo.Phone);
    MySqlParameter p4 = new MySqlParameter("@certType", checkInRoomInfo
    .CertType);
    MySqlParameter p5 = new MySqlParameter("@certId", checkInRoomInfo.CertId);
    MySqlParameter p6 = new MySqlParameter("@address", checkInRoomInfo
    .Address);
    MySqlParameter p7 = new MySqlParameter("@personNum", checkInRoomInfo
    .PersonNum);
    MySqlParameter p8 = new MySqlParameter("@inId", checkInRoomInfo.InId);
    MySqlParameter[] paramArray = new MySqlParameter[] { p1, p2, p3, p4, p5,
    p6, p7, p8 };
    if (DBOper.ExecuteCommand(sqlUpdateRegister, paramArray) == 1)
    {
        return true;
```

```
        }
        else
        {
            return false;
        }
    }
}
```

【分析】

在住客信息查询界面中,操作员可以将查询到的住客姓名等信息进行修改并保存在 checkin(入住信息表)中。

(4) GetClient()方法。用于从 checkin 表中根据房间号 roomId 查找住客信息,返回值为 DataTable 类型,程序代码如下:

```
public static DataTable GetClient(string roomId)
{
    string sql = @"select roomId as 房间号,price as 价格,foregift as 押金,
        inTime as 入住时间,outTime as 退房时间,clientName as 住客姓名, gender as 性
            别,phone as 电话号码,certType as 证件类型,certId as 证件号码,address as
            地址,personNum as 入住人数, oper as 登记者 from checkin where delMark=0
            and roomId =" + roomId;
    DataTable dt = DBOper.GetDataTable(sql);
    return dt;
}
```

【分析】

在客户信息查询界面中,需要从 checkin(入住信息表)中根据用户选择的房间号 roomId 查询出住客信息。

8.7 表示层设计

酒店管理系统中的模块和对应的窗体名称如表 8-5 所示。

表 8-5 酒店管理系统窗体描述表

模 块 名	窗 体	窗体 Name
登录	登录	LoginForm
主窗体	系统主界面	MainForm
用户管理	添加新用户	AddUserForm
	修改/删除用户	UserManagerForm
客房管理	入住登记	CheckInRoomForm
	退房登记	CheckOutRoomForm
查询管理	住客信息查询	ClientForm
	客房信息查询	RoomForm
帮助	关于	HelpAbout

8.7.1 登录设计

酒店管理系统的"登录"窗体是首界面,登录信息包括用户名和密码,用于实现对用户的身份验证。

1. 窗体设计

"登录"窗体如图 8-2 所示。

图 8-2 "登录"窗体

"登录"窗体和控件的属性设置如表 8-6 所示。

表 8-6 "登录"窗体和控件的属性设置

Name	类 型	Text 属性	其他属性
LoginForm	Form	酒店管理系统	ControlBox:True StartPosition:CenterScreen
pictureBox1	PictureBox		Image 设置为选定的图片 SizeMode:StretchImage
groupBox1	GroupBox	登录	
label1	Label	用户名	
label2	Label	密码	
txtName	TextBox		
txtPwd	TextBox		UseSystemPasswordChar:True
btnLogin	Button	登录系统	
btnExit	Button	退出系统	

2. 功能实现

(1) 登录系统。用户输入用户名和密码后,单击"登录系统"按钮后,如果登录成功,则直接进入系统主界面；如果登录失败，则显示提示信息"用户名或密码错误"。"登录系统"的 Click 事件处理程序代码如下：

```csharp
private void btnLogin_Click(object sender, EventArgs e)
{
    string strName = txtName.Text;
    string strPwd = txtPwd.Text;
    if (string.IsNullOrEmpty(strName))
    {
        MessageBox.Show("用户名不能为空", "提示", MessageBoxButtons.OK,
        MessageBoxIcon.Information);
        return;
    }
    if (string.IsNullOrEmpty(strPwd))
    {
        MessageBox.Show("密码不能为空",提示", MessageBoxButtons.OK,
        MessageBoxIcon. Information);
        return;
    }
    User user = UserManager.GetUser(strName, strPwd);
    if (user == null)
    {
        MessageBox.Show("用户名或密码错误", "提示", MessageBoxButtons. OK,
        MessageBoxIcon.Information);
    }
    else
    {
        RoleManager.curUser = user;
        this.Hide();
        MainForm mainform = new MainForm();
        mainform.Show();
    }
}
```

（2）退出系统。用户单击"退出系统"按钮，直接退出酒店管理系统。"退出系统"的 Click 事件处理程序代码如下：

```csharp
private void btnExit_Click(object sender, EventArgs e)
{
    Application.Exit();
}
```

8.7.2 系统主界面设计

酒店管理系统的主界面采用 MDI 界面设计，主要用于导航。

1. 窗体设计

主界面如图 8-3 所示。

图 8-3 主界面

主界面窗体和控件的属性设置如表 8-7 所示。

表 8-7 主界面窗体和控件的属性设置

Name	类 型	Text 属性	其 他 属 性
MainForm	Form	酒店管理系统	WindowState：Maximized IsMdiContainer：True
menuStrip1	MenuStrip	设为相应的文本	ToolStripDropDownMenu： miCheckIn miCheckOut miRoom miClient miNewUser miUpdateUser miAbout
toolStrip1	ToolStrip	设为相应的文本	ToolStripButton： DisplayStyle：ImageAndText Image：设为相应的图标 TextImageRelation：ImageBeforeText

2．功能实现

（1）主界面加载。根据用户角色显示相应的菜单和工具栏，如果为员工用户，即 Role 属性值为 0 时，用户管理相关功能不可用。主界面的 Load 事件处理程序代码如下：

```
private void MainForm_Load(object sender, EventArgs e)
{
    if (RoleManager.curUser.Role == 0)
    {
        miUserManage.Visible = false;
        miNewUser.Visible = false;
        miEditUser.Visible = false;
    }
}
```

（2）入住登记。单击"入住登记"菜单或工具栏，显示入住登记窗体。"入住登记"的 Click 事件处理程序代码如下：

```csharp
private void miCheckIn_Click(object sender, EventArgs e)
{
    CheckInRoomForm frm = new CheckInRoomForm();
    frm.MdiParent = this;
    frm.Show();
}
```

（3）退房登记。单击"退房登记"菜单或工具栏，显示退房登记窗体。"退房登记"的 Click 事件处理程序代码如下：

```csharp
private void miCheckOut_Click(object sender, EventArgs e)
{
    CheckOutRoomForm frm = new CheckOutRoomForm();
    frm.MdiParent = this;
    frm.Show();
}
```

（4）住客信息。单击"住客信息"菜单或工具栏，显示住客信息查询窗体。"住客信息"的 Click 事件处理程序代码如下：

```csharp
private void miClientQuery_Click(object sender, EventArgs e)
{
    ClientForm frm = new ClientForm();
    frm.MdiParent = this;
    frm.Show();
}
```

（5）客房信息。单击"客房信息"菜单或工具栏，显示客房信息查询窗体。"客房信息"的 Click 事件处理程序代码如下：

```csharp
private void miRoomQuery_Click(object sender, EventArgs e)
{
    RoomForm frm = new RoomForm();
    frm.MdiParent = this;
    frm.Show();
}
```

（6）添加新用户。单击"添加新用户"菜单或工具栏，显示添加新用户窗体。"添加新用户"的 Click 事件处理程序代码如下：

```csharp
private void miNewUser_Click(object sender, EventArgs e)
{
    AddUserForm frm = new AddUserForm();
    frm.MdiParent = this;
    frm.Show();
}
```

（7）修改/删除用户。单击"修改/删除用户"菜单或工具栏，显示修改/删除用户窗体。"修改/删除用户"的 Click 事件处理程序代码如下：

```csharp
private void miEditUser_Click(object sender, EventArgs e)
{
    UserManagerForm frm = new UserManagerForm();
```

```
        frm.MdiParent = this;
        frm.Show();
    }
```

（8）关于。单击"关于"菜单，显示关于窗体。"关于"的 Click 事件处理程序代码如下：

```
private void miAbout_Click(object sender, EventArgs e)
{
    HelpAbout frm = new HelpAbout();
    frm.MdiParent = this;
    frm.Show();
}
```

8.7.3 添加新用户设计

酒店管理系统的"添加新用户"窗体，主要用于添加管理员或员工的用户名、密码和权限。

1. 窗体设计

"添加新用户"窗体如图 8-4 所示。

图 8-4 "添加新用户"窗体

"添加新用户"窗体和控件的属性设置如表 8-8 所示。

表 8-8 "添加新用户"窗体和控件的属性设置

Name	类型	Text 属性	其他属性
AddUserForm	Form	添加新用户	
label1	Label	用户名	
label2	Label	密码	
label3	Label	权限	
txtName	TextBox		
txtPwd	TextBox		UseSystemPasswordChar：True
rbEmp	RadioButton	员工	Checked：True
rbAdmin	RadioButton	管理员	
btnAdd	Button	添加	
btnCancel	Button	取消	

2. 功能实现

（1）添加。用户在文档框中输入新用户的用户名和密码，并且选择用户角色，单击"添加"按钮，将新用户信息保存到数据库 user 表中。"添加"的 Click 事件处理程序代码如下：

```
private void btnAdd_Click(object sender, EventArgs e)
{
    string strName = txtName.Text;
    string strPwd = txtPwd.Text;
    int role = 0;
    if (rbAdmin.Checked)
    {
        role = 1;
    }
    if (string.IsNullOrEmpty(strName))
    {
        MessageBox.Show("用户名不能为空", "提示", MessageBoxButtons.OK,
        MessageBoxIcon.Information);
        txtName.Focus();
        return;
    }
    if (string.IsNullOrEmpty(strPwd))
    {
        MessageBox.Show("密码不能为空", "提示", MessageBoxButtons.OK,
        MessageBoxIcon.Information);
        txtPwd.Focus();
        return;
    }
    User user = new User(strName,strPwd,role);
    if (UserManager.AddUser(user))
    {
        MessageBox.Show("添加成功");
    }
    else
    {
        MessageBox.Show("添加失败");
    }
}
```

（2）取消。用户单击"取消"按钮，将关闭当前窗体。"取消"的 Click 事件处理程序代码如下：

```
private void btnCancel_Click(object sender, EventArgs e)
{
    this.Close();
}
```

8.7.4 修改/删除用户设计

酒店管理系统的"修改/删除用户"窗体,主要用于修改或删除用户。

1. 窗体设计

"修改/删除用户"窗体如图 8-5 所示。

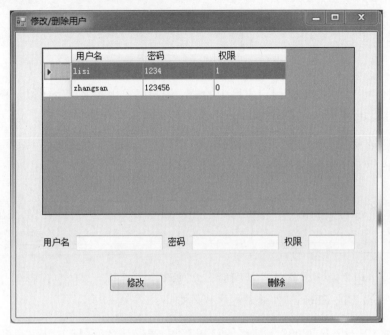

图 8-5 "修改/删除用户"窗体

"修改/删除用户"窗体和控件的属性设置如表 8-9 所示。

表 8-9 "修改/删除用户"窗体和控件的属性设置

Name	类型	Text 属性	其他属性
UserManagerForm	Form	修改/删除用户	ControlBox: Ture StartPosition: CenterScreen
dgvUsers	DataGridView		AllowUserToAddRows: False AllowUserToDeleteRows: False ReadOnly: True SelectionMode: FullRowSelect MultiSelect: False
txtUserName	TextBox		
txtPwd	TextBox		
txtRole	TextBox		
btnChange	Button	修改	
btnDel	Button	删除	

2. 功能实现

（1）修改/删除用户界面加载。界面加载时，将在 dgvUsers 中显示所有用户信息，dgvUsers 控件中的列的 HeaderText 属性分别为用户名、密码、权限，采用设置数据源的方式进行数据绑定。修改/删除用户界面的 Load 事件处理程序代码如下：

```
private void UserManagerForm_Load(object sender, EventArgs e)
{
    DataTable dt = UserManager.GetUserInfo();
    dgvUsers.DataSource = dt;
}
```

（2）选择用户。用户单击用户表格中的某个用户时，用户名、密码以及权限将显示到相应文本框中。dgvUsers 的 CellContentClick 事件处理程序代码如下：

```
private void dgvUsers_CellContentClick(object sender,
DataGridViewCellEventArgs e)
{
    txtUserName.Text = dgvUsers.CurrentRow.Cells[0].Value.ToString();
    txtPwd.Text = dgvUsers.CurrentRow.Cells[1].Value.ToString();
    txtRole.Text = dgvUsers.CurrentRow.Cells[2].Value.ToString();
}
```

（3）修改。用户单击"修改"按钮时，实现将修改后的用户信息保存到 hotel 数据库 user 表中。"修改"的 Click 事件处理程序代码如下：

```
private void btnChange_Click(object sender, EventArgs e)
{
    User user = new User();
    user.Name = txtUserName.Text;
    user.Pwd = txtPwd.Text;
    user.Role = int.Parse(txtRole.Text);
    if (UserManager.UpdateUser(user))
    {
        MessageBox.Show("用户信息修改成功");
        DataTable dt = UserManager.GetUserInfo();
        dgvUsers.DataSource = dt;
    }
    else
    {
        MessageBox.Show("用户信息修改失败");
    }
}
```

（4）删除。用户单击"删除"按钮时，实现将 Hotel 数据库 user 表中的用户信息删除。"删除"的 Click 事件处理程序代码如下：

```
private void btnDel_Click(object sender, EventArgs e)
{
```

```
        if (MessageBox.Show("确定要删除用户数据吗？", "提示",
MessageBoxButtons.OKCancel, MessageBoxIcon.Information) == DialogResult.OK)
        {
            User user = new User();
            user.Name = txtUserName.Text;
            user.Pwd = txtPwd.Text;
            user.Role = int.Parse(txtRole.Text);
            if (UserManager.DeleteUser(user))
            {
                MessageBox.Show("用户删除成功");
                DataTable dt = UserManager.GetUserInfo();
                dgvUsers.DataSource = dt;
            }
            else
            {
                MessageBox.Show("用户删除失败");
            }
        }
    }
```

8.7.5 入住登记设计

酒店管理系统的"入住登记"窗体，主要用于选择房间和登记住客的个人信息。

1. 窗体设计

"入住登记"窗体如图 8-6 所示。

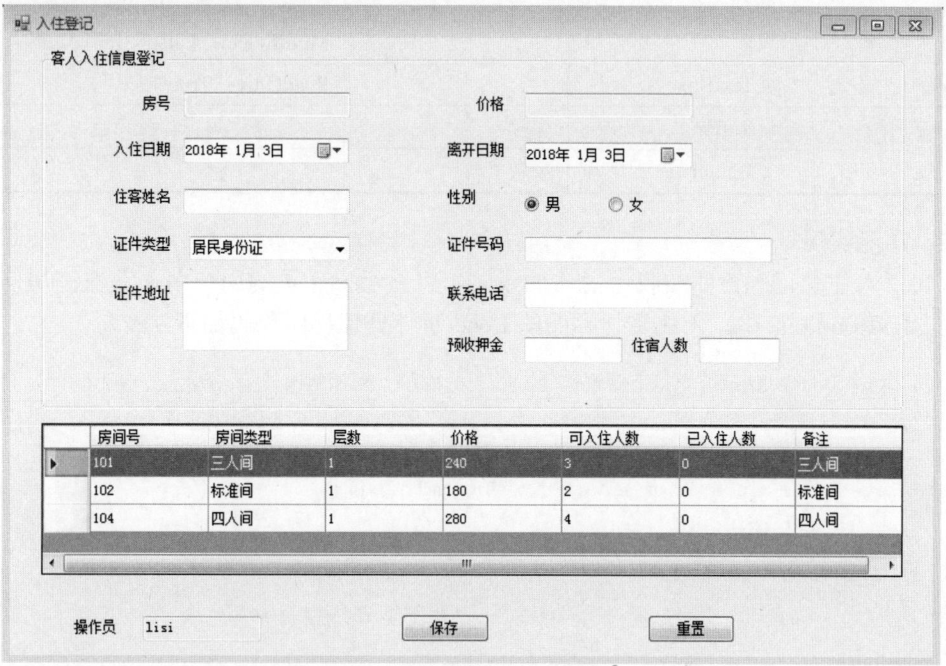

图 8-6 "入住登记"窗体

"入住登记"窗体和控件的属性设置如表 8-10 所示。

表 8-10 "入住登记"窗体和控件的属性设置

Name	类 型	Text 属性	其 他 属 性
CheckInRoomForm	Form	入住登记	
groupBox1	GroupBox	客人入住信息登记	
txtRoomId	TextBox		ReadOnly：True
txtPrice	TextBox		ReadOnlyy：True
dtpInTime	DateTimePicker		
dtpOutTime	DateTimePicker		
txtClientName	TextBox		
rbMale	RadioButtom	男	Checkedy：True
rbFemale	RadioButtom	女	
cmbCertType	ComboBox	居民身份证	Itemsy：居民身份证、学生证、工作证
txtCertId	TextBox		
txtAddress	TextBox		Multiline：True
txtPhone	TextBox		
txtForegift	TextBox		
txtPersonNum	TextBox		
dgvRoomList	DataGridView		AllowUserToAddRows：False AllowUserToDeleteRows：False ReadOnly：True SelectionMode：FullRowSelect MultiSelect：False
txtOper	TextBox		ReadOnly：True
btnSave	Button	保存	
btnReset	Button	重置	

2．功能实现

（1）"入住登记"窗体加载。窗体加载时，将显示操作员的用户名，获取空房信息并显示在 dgvRoomList 中。入住登记界面的 Load 事件处理程序代码如下：

```
private void CheckInRoomForm_Load(object sender, EventArgs e)
{
    txtOper.Text = RoleManager.curUser.Name;
    DataTable dt = RoomManager.GetRoomInfo();
    if (dt.Rows.Count!=0)
    {
        dgvRoomList.DataSource = dt;
    }
    else
    {
```

```
        MessageBox.Show("已没有空房间!");
        return;
    }
}
```

（2）选择房间。用户单击房间表格中的某个房间时，房间号与价格将显示到相应文本框中。dgvRoomList 的 CellContentClick 事件处理程序代码如下：

```
private void dgvRoomList_CellContentClick(object sender, DataGridViewCell-
EventArgs e)
{
    if (dgvRoomList.SelectedRows.Count == 1)
    {
        this.txtRoomId.Text = dgvRoomList.Selected-Rows[0]
.Cells["房间号"].Value.ToString();
        this.txtPrice.Text = dgvRoomList.SelectedRows[0]
.Cells["价格"].Value.ToString();
    }
}
```

（3）验证用户输入。封装 CheckData()方法，实现验证用户输入的合法性，其中房间号不能为空，价格不能为空，押金不能为空，住客姓名不能为空，电话号不能为空，证件号不能为空，地址不能为空，住宿人数不能为空且不能超过房间可容纳人数，入住时间不能小于当前时间，离开时间不能小于当前时间，并且离开时间不能小于入住时间。CheckData()方法程序代码如下：

```
private bool CheckData()
{
    if (txtRoomId.Text == "")
    {
        MessageBox.Show("房间号不能为空");
        txtRoomId.Focus();
        return false;
    }
    if (txtPrice .Text == "")
    {
        MessageBox.Show("价格不能为空");
        txtPrice.Focus();
        return false;
    }
    if (txtForegift .Text == "")
    {
        MessageBox.Show("押金不能为空");
        txtForegift.Focus();
        return false;
    }
    if (txtClientName.Text == "")
    {
```

```csharp
        MessageBox.Show("住客姓名不能为空");
        txtClientName.Focus();
        return false;
    }
    if (txtPhone.Text == "")
    {
        MessageBox.Show("电话不能为空");
        txtPhone.Focus();
        return false;
    }
    if (txtCertId.Text == "")
    {
        MessageBox.Show("证件号不能为空");
        txtCertId.Focus();
        return false;
    }
    if(txtAddress.Text == "")
    {
        MessageBox.Show("地址不能为空");
        txtAddress.Focus();
        return false;
    }
    if (txtPersonNum.Text == "")
    {
        MessageBox.Show("住宿人数不能为空");
        txtPersonNum.Focus();
        return false;
    }
    else if (Convert.ToInt32(txtPersonNum.Text) > Convert.ToInt32
    (dgvRoomList.SelectedRows[0].Cells["可容纳人数"].Value.ToString()))
    {
        MessageBox.Show("入住人数超过可容纳人数,请更换房间");
        txtRoomId.Text = "";
        txtPrice.Text = "";
        txtPersonNum.Text = "";
        txtRoomId.Focus();
        return false;
    }
    DateTime inTime = DateTime.Parse(dtpInTime.Text.ToString());
    DateTime outTime = DateTime.Parse(dtpOutTime.Text.ToString());
    if(inTime.CompareTo(DateTime.Today)<0)
    {
        MessageBox.Show("入住时间不能小于当前时间");
        dtpInTime.Focus();
        return false;
    }
    else if(outTime.CompareTo(DateTime.Today)<0)
    {
        MessageBox.Show("离开时间不能小于当前时间");
```

```csharp
            dtpOutTime.Focus();
            return false;
        }
        else if(outTime < inTime)
        {
            MessageBox.Show("离开时间不能小于入住时间");
            dtpOutTime.Focus();
            return false;
        }
        return true;
    }
```

(4) 保存。用户单击"保存"按钮时，验证用户输入合法后，实现将入住信息保存到 hotel 数据库 checkin 表中。"保存"的 Click 事件处理程序代码如下：

```csharp
    private void btnSave_Click(object sender, EventArgs e)
    {
        string strTimeNow = string.Format("{0:T}",DateTime.Now);
        string inTime = dtpInTime.Text.ToString() + strTimeNow;
        string outTime = dtpOutTime.Text.ToString() + strTimeNow;
        int gender = 1;
        if (rbMale.Checked)
        {
            gender = 0;
        }
        CheckInRoom registerInfo = new CheckInRoom();
        registerInfo.RoomId = txtRoomId.Text;
        registerInfo.Price = Convert.ToDouble(txtPrice.Text);
        registerInfo.Foregift = Convert.ToDouble(txtForegift.Text);
        registerInfo.InTime = Convert.ToDateTime(inTime);
        registerInfo .OutTime = Convert.ToDateTime(outTime);
        registerInfo .ClientName = txtClientName.Text;
        registerInfo .Gender = gender;
        registerInfo .Phone = txtPhone.Text;
        registerInfo .CertType = cmbCertType.SelectedItem.ToString();
        registerInfo.CertId = txtCertId.Text;
        registerInfo.PersonNum = Convert.ToInt32(txtPersonNum.Text);
        registerInfo.Oper = txtOper.Text;
        registerInfo.Address = txtAddress.Text;
        registerInfo.DelMark = 0;
        if (CheckData())
        {
            if (RoomManager.InsertRoomInfo(registerInfo))
            {
                MessageBox.Show("入住登记成功!");
            }
            else
            {
```

```
            MessageBox.Show("入住登记失败！");
        }
    }
}
```

（5）重置。用户单击"重置"按钮时，将清空文本框等控件的数据。"重置"的 Click 事件处理程序代码如下：

```
private void btnReset_Click(object sender, EventArgs e)
{
    txtRoomId.Text = "";
    txtPrice.Text = "";
    txtForegift.Text = "";
    dtpInTime.Text = "";
    dtpOutTime.Text = "";
    rbMale.Checked = true;
    txtClientName.Text = "";
    txtPhone.Text = "";
    cmbCertType.SelectedIndex = 0;
    txtCertId.Text = "";
    txtAddress.Text = "";
    txtPersonNum.Text = "";
}
```

8.7.6 退房登记设计

酒店管理系统的"退房登记"窗体，主要用于计算住宿费用和登记住客退房信息。

1. 窗体设计

"退房登记"窗体如图 8-7 所示。

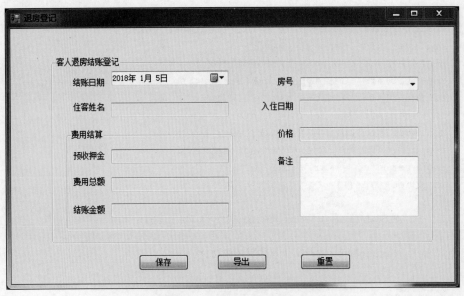

图 8-7 "退房登记"窗体

"退房登记"窗体和控件的属性设置如表 8-11 所示。

表 8-11 "退房登记"窗体和控件的属性设置

Name	类型	Text 属性	其他属性
CheckOutRoomForm	Form	退房登记	
groupBox1	GroupBox	客人退房结账登记	
dtpOutTime	DateTimePicker		
cmbRoomId	ComboBox		
txtClientName	TextBox		ReadOnly：True
txtInTime	TextBox		ReadOnly：True
txtPrice	TextBox		ReadOnly：True
groupBox2	GroupBox	费用结算	
txtForegift	TextBox		ReadOnly：True
txtTotal	TextBox		ReadOnly：True
txtAccount	TextBox		ReadOnly：True
txtNote	TextBox		Mutiline：True
btnSave	Button	保存	
btnPrint	Button	导出	
btnReset	Button	重置	

2．功能实现

（1）获取入住房间信息。在 CheckOutRoomForm 窗体初始化时，获取已入住房间信息。CheckOutRoomForm 的构造方法程序代码如下：

```
public partial class CheckOutRoomForm : Form
{
    private DataTable dt;
    public CheckOutRoomForm()
    {
        dt = RoomManager.GetUseRoomInfo();
        InitializeComponent();
    }
}
```

（2）显示房间号。在 CheckOutRoomForm 窗体的 Load 事件中，获取房间号，并在 cmbRoomId 中显示。"退房登记"窗体的 Load 事件处理程序代码如下：

```
private void CheckOutRoomForm_Load(object sender, EventArgs e)
{
    if (dt.Rows.Count > 0)
    {
        for (int i = 0; i <= dt.Rows.Count - 1; i++)
        {
            cmbRoomId.Items.Add(dt.Rows[i]["房间号"].ToString());
        }
    }
}
```

（3）选择退房房间。当用户在 cmbRoomId 中选择某一房间号时，该房间的相关住客信息在相应的文本框等控件中显示，并且计算和显示住宿费用。cmbRoomId 的 SelectedIndexChanged 事件处理程序代码如下：

```csharp
private void cmbRoomId_SelectedIndexChanged(object sender, EventArgs e)
{
    for (int i = 0; i < dt.Rows.Count; i++)
    {
        if (cmbRoomId.SelectedItem.ToString() == dt.Rows[i]["房间号"]
        .ToString())
        {
            txtPrice.Text = dt.Rows[i]["价格"].ToString();
            txtForegift.Text = dt.Rows[i]["押金"].ToString();
            txtInTime.Text = dt.Rows[i]["入住时间"].ToString();
            txtClientName.Text = dt.Rows[i]["住客姓名"].ToString();
            string outTime = dtpOutTime.Text + string.Format("{0:T}",
            DateTime.Now);
            TimeSpan ts1 = new TimeSpan(DateTime.Parse(outTime).Ticks);
            TimeSpan ts2 = new TimeSpan(DateTime.Parse(txtInTime.Text)
            .Ticks);
            TimeSpan ts = ts1.Subtract(ts2);
            int dayCount = ts.Days;
            int hourCount = ts.Hours;
            if (dayCount == 0 & hourCount < 24)
            {
                dayCount = 1;
            }
            double consumTotal = double.Parse(txtPrice.Text.ToString()) *
            dayCount;
            txtTotal.Text = consumTotal.ToString();
            txtAccount.Text = Convert.ToString(consumTotal - double
            .Parse(txtForegift.Text));
            return;
        }
    }
}
```

（4）验证。封装 checkdata()方法，用于在保存退房信息时进行数据验证，其中结账日期不能小于入住时期。checkdata()方法程序代码如下：

```csharp
private bool checkdata()
{
    DateTime inTime = Convert.ToDateTime(DateTime.Parse(txtInTime
    .Text.ToString()).ToString("yyyy-MM-dd"));
    DateTime outTime = DateTime.Parse(dtpOutTime.Text.ToString());
    if (outTime < inTime)
    {
        MessageBox.Show("结账日期不能小于入住日期");
```

```
        dtpOutTime.Focus();
        return false;
    }
    return true;
}
```

(5) 保存。用户单击"保存"按钮时,验证用户输入合法后,实现将入住信息保存到 hotel 数据库 checkout 表中。"保存"的 Click 事件处理程序代码如下:

```
private void btnSave_Click(object sender, EventArgs e)
{
    if (checkdata() == false)
    {
        return;
    }
    string inTime = txtInTime.Text.ToString();
    string strTimeNow = string.Format("{0:T}", DateTime.Now);
    string outTime = dtpOutTime.Text.ToString() + strTimeNow;
    int inId = int.Parse(RoomManager.GetInID(cmbRoomId.SelectedItem
    .ToString()).Rows[0][0].ToString());
    CheckOutRoom checkOutRoom = new CheckOutRoom();
    checkOutRoom.OutId = DateTime.Now.ToString("yyyyMMddHHmmss");
    checkOutRoom.InId = inId;
    checkOutRoom.RoomId = cmbRoomId.SelectedItem.ToString();
    checkOutRoom.Price = double.Parse(txtPrice.Text.ToString());
    checkOutRoom.Foregift = double.Parse(txtForegift.Text.ToString());
    checkOutRoom.Total = double.Parse(txtTotal.Text.ToString());
    checkOutRoom.Account = double.Parse(txtAccount.Text.ToString());
    checkOutRoom.InTime = DateTime.Parse(inTime);
    checkOutRoom.OutTime = DateTime.Parse(outTime);
    checkOutRoom.ClientName = txtClientName.Text.ToString();
    checkOutRoom.Oper = RoleManager.curUser.Name;
    checkOutRoom.Note = txtNote.Text.ToString();
    if (RoomManager.InsertCheckOutRoomInfo(checkOutRoom))
    {
        MessageBox.Show("退房信息保存成功");
        return;
    }
    else
    {
        MessageBox.Show("退房信息保存失败");
        return;
    }
}
```

(6) 导出。用户单击"导出"按钮时,将住客的发票单导出到 D 盘发票单文件夹中,发票单以时间命名。"导出"的 Click 事件处理程序代码如下:

```
private void btnPrint_Click(object sender, EventArgs e)
```

```
{
    string path = "D:\\发票单\\" + DateTime.Now.ToString("yyyyMMddHHmmss")
        + ".txt";
    StreamWriter w = new StreamWriter(path);
    w.WriteLine("--专家公寓--");
    w.WriteLine("住客姓名: " + txtClientName.Text);
    w.WriteLine("住宿费: " + txtTotal.Text);
    w.WriteLine("开票日期: " + DateTime.Now.ToString("yyyy-MM-dd"));
    w.Close();
    MessageBox.Show("导出成功");
}
```

（7）重置。用户单击"重置"按钮时，将清空文本框等控件的数据。"重置"的 Click 事件处理程序代码如下：

```
private void btnReset_Click(object sender, EventArgs e)
{
    cmbRoomId.Text = "";
    txtClientName.Text = "";
    txtInTime.Text = "";
    txtPrice.Text = "";
    txtForegift.Text = "";
    txtTotal.Text = "";
    txtAccount.Text = "";
    txtNote.Text = "";
}
```

8.7.7 住客信息查询设计

酒店管理系统的"住客信息"窗体，主要用于对住客信息的查询和修改。

1．窗体设计

"住客信息"窗体如图 8-8 所示。

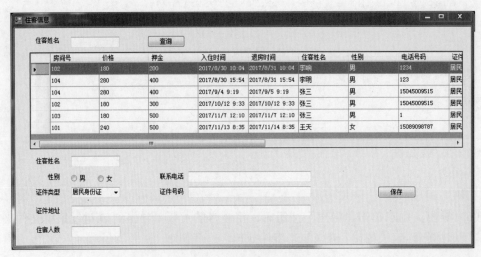

图 8-8 "住客信息"窗体

"住客信息"窗体和控件的属性设置如表8-12所示。

表8-12 "住客信息"窗体和控件的属性设置

Name	类型	Text属性	其他属性
ClientForm	Form	住客信息	
txtQueryName	TextBox		
btnQuery	Button	查询	
dgvClient	DataGridView		AllowUserToAddRows: False AllowUserToDeleteRows: False ReadOnly: True SelectionMode: FullRowSelect MultiSelect: False
txtClientName	TextBox		
rbMale	RadioButton	男	Checked: True
rbFemale	RadioButton	女	
txtPhone	TextBox		
cmbCertType	ComboBox	居民身份证	Items: 居民身份证、学生证、工作证
txtCertId	TextBox		
txtAddress	TextBox		
txtPersonNum	TextBox		
btnSave	Button	保存	

2. 功能实现

（1）"住客信息"窗体界面加载。窗体加载时，将住客信息显示在dgvClient中。住客信息查询界面的Load事件处理程序代码如下：

```
private void ClientForm_Load(object sender, EventArgs e)
{
    DataTable dt = ClientManager.GetClientInfo();
    dgvClient.DataSource = dt;
}
```

（2）格式化。在dgvClient显示数据时，将性别以及是否退房数据进行格式化显示。dgvClient的CellFormatting事件处理程序代码如下：

```
private void dgvClient_CellFormatting(object sender, DataGridViewCell-
FormattingEventArgs e)
{
    if (e.ColumnIndex == 6)
    {
        if (e.Value.ToString() == "1")
        {
```

```
                e.Value = "女";
            }
            else
            {
                e.Value = "男";
            }
        }
        if (e.ColumnIndex == 13)
        {
            if (e.Value.ToString() == "0")
            {
                e.Value = "否";
            }
            else
            {
                e.Value = "是";
            }
        }
    }
```

(3) 查询。根据用户输入的住客姓名进行模糊查询住客信息。btnQuery 的 Click 事件处理程序代码如下：

```
private void btnQuery_Click(object sender, EventArgs e)
{
    string clientName = txtQueryName.Text;
    DataTable dt = ClientManager.GetALLClient(clientName);
    dgvClient.DataSource = dt;
}
```

(4) 选择住客。当用户在 dgvClient 中选择某一住客时，该住客的相关信息在相应的文本框等控件中显示。dgvClient 的 CellContentClick 事件处理程序代码如下：

```
private void dgvClient_CellContentClick(object sender, DataGridViewCell-
EventArgs e)
{
    var rows = dgvClient.SelectedRows;
    if (rows.Count == 0)
    {
        txtClientName.Text = "";
        rbMale.Checked = false;
        rbFemale.Checked = false;
        txtPhone.Text = "";
        cmbCertType.SelectedItem = "";
```

```
        txtCertId.Text = "";
        txtAddress.Text = "";
        txtPersonNum.Text = "";
        btnSave.Enabled = false;
        return;
    }
    btnSave.Enabled = true;
    var row = rows[0];
    txtClientName.Text = row.Cells[5].Value.ToString();
    if (row.Cells[6].Value.ToString() == "0")
    {
        rbMale.Checked = true;
    }
    else
    {
        rbFemale.Checked = true;
    }
    txtPhone.Text = row.Cells[7].Value.ToString();
    cmbCertType.SelectedItem = row.Cells[8].Value;
    txtCertId.Text = row.Cells[9].Value.ToString();
    txtAddress.Text = row.Cells[10].Value.ToString();
    txtPersonNum.Text = row.Cells[11].Value.ToString();
}
```

（5）验证。封装 checkdata()方法，用于在保存修改后的住客信息时进行数据验证。checkdata()方法程序代码如下：

```
private bool checkdata()
{
    if (txtClientName.Text == "")
    {
        MessageBox.Show("住客姓名不能为空");
        txtClientName.Focus();
        return false;
    }
    if (txtPhone.Text == "")
    {
        MessageBox.Show("电话号不能为空");
        txtPhone.Focus();
        return false;
    }
    if (txtCertId.Text == "")
    {
```

```csharp
        MessageBox.Show("证件号不能为空");
        txtCertId.Focus();
        return false;
    }
    if (txtAddress.Text == "")
    {
        MessageBox.Show("地址不能为空");
        txtAddress.Focus();
        return false;
    }
    DataTable dt = RoomManager.GetRoomInfo(clientGridView.SelectedRows[0]
    .Cells[0].Value.ToString());
    string personNum = dt.Rows[0][4].ToString();
    if (txtPersonNum.Text == "")
    {
        MessageBox.Show("住宿人数不能为空");
        txtPersonNum.Focus();
        return false;
    }
    else if (int.Parse(txtPersonNum.Text) > int.Parse(personNum))
    {
        MessageBox.Show("入住人数大于可容纳人数,请重新输入!");
        return false;
    }
    return true;
}
```

（6）保存。用户单击"保存"按钮时，验证用户输入合法后，实现将住客信息保存到 hotel 数据库 checkin 表中。保存的 Click 事件处理程序代码如下：

```csharp
private void btnSave_Click(object sender, EventArgs e)
{
    int gender;
    if (rbMale.Checked)
    {
        gender = 0;
    }
    else
    {
        gender = 1;
    }
    CheckInRoom registerInfo = new CheckInRoom();
```

```csharp
registerInfo.RoomId = dgvClient.SelectedRows[0].Cells[0].Value.ToString();
registerInfo.Price = double.Parse(dgvClient.SelectedRows[0].Cells[1].Value.ToString());
registerInfo.Foregift = double.Parse(dgvClient.SelectedRows[0].Cells[2].Value.ToString());
registerInfo.InTime = DateTime.Parse(dgvClient.SelectedRows[0].Cells[3].Value.ToString());
registerInfo.OutTime = DateTime.Parse(dgvClient.SelectedRows[0].Cells[4].Value.ToString());
registerInfo.ClientName = txtClientName.Text.ToString();
registerInfo.Gender = gender;
registerInfo.Phone = txtPhone.Text.ToString();
registerInfo.CertType = cmbCertType.SelectedItem.ToString();
registerInfo.CertId = txtCertId.Text.ToString();
registerInfo.PersonNum = int.Parse(txtPersonNum.Text.ToString());
registerInfo.Address = txtAddress.Text.ToString();
registerInfo.Oper = dgvClient.SelectedRows[0].Cells[12].Value.ToString();
registerInfo.DelMark = int.Parse(dgvClient.SelectedRows[0].Cells[13].Value.ToString());
registerInfo.InId = int.Parse(dgvClient.SelectedRows[0].Cells[14].Value.ToString());
if (checkdata())
{
    if (ClientManager.UpdateRegister(registerInfo))
    {
    MessageBox.Show("住客信息修改成功");
    DataTable dt = ClientManager.GetClientInfo();
    dgvClient.DataSource = dt;
    }
    else
    {
        MessageBox.Show("住客信息修改失败");
    }
}
}
```

8.7.8 客房信息查询设计

酒店管理系统的"客房信息"窗体，主要用于房间信息进行查询。

1. 窗体设计

"客房信息"窗体如图 8-9 所示。

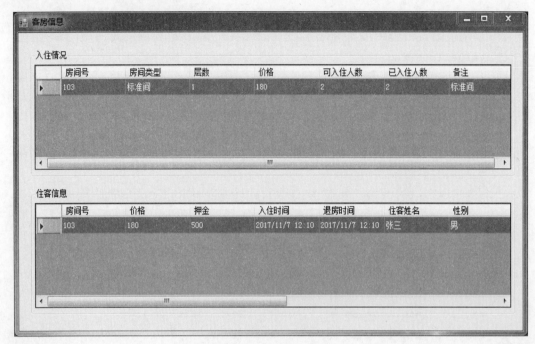

图 8-9 "客房信息"窗体

"客房信息"窗体和控件的属性设置如表 8-13 所示。

表 8-13 "客房信息"窗体和控件的属性设置

Name	类型	Text 属性	其他属性
RoomForm	Form	客房信息	
groupBox1	GroupBox	入住情况	
dgvRoom	DataGridView		AllowUserToAddRows：False AllowUserToDeleteRows：False ReadOnly：True SelectionMode：FullRowSelect MultiSelect：False
groupBox2	GroupBox	住客信息	
dgvClient	DataGridView		AllowUserToAddRows：False AllowUserToDeleteRows：False ReadOnly：True SelectionMode：FullRowSelect MultiSelect：False

2．功能实现

（1）"客房信息"窗体加载。窗体加载时，将客房信息显示在 dgvRoom 中。客房信息查询界面的 Load 事件处理程序代码如下：

```
private void RoomForm_Load(object sender, EventArgs e)
{
```

```
        DataTable dt = RoomManager.GetRoomUseInfo();
        dgvRoom.DataSource = dt;
}
```

(2)选择客房。用户在 **dgvRoom** 中选择某一客房时，该客房的相关信息在相应的文本框等控件中显示。dgvRoom 的 SelectionChanged 事件处理程序代码如下：

```
private void dgvRoom_SelectionChanged(object sender, EventArgs e)
{
    if (dgvRoom.SelectedRows.Count != 0)
    {
        string roomId = dgvRoom.SelectedRows[0].Cells[0].Value.ToString();
        DataTable dt = ClientManager.GetClient(roomId);
        dgvClient.DataSource = dt;
    }
}
```

(3)格式化。在 dgvClient 显示数据时，将性别等数据进行格式化显示。dgvClient 的 CellFormatting 事件处理程序代码如下：

```
private void dgvClient_CellFormatting(object sender, DataGridViewCell-
FormattingEventArgs e)
{
    if(e.ColumnIndex == 6)
    {
        if (e.Value.ToString() == "1")
        {
            e.Value = "女";
        }
        else
        {
            e.Value = "男";
        }
    }
}
```

8.7.9 帮助设计

系统的帮助模块是一个"关于"框，用于提示系统版本及版权等信息。

1. 窗体设计

"关于"窗体界面如图 8-10 所示，使用 Visual Studio 中提供的"关于"框窗体模板进行设计。

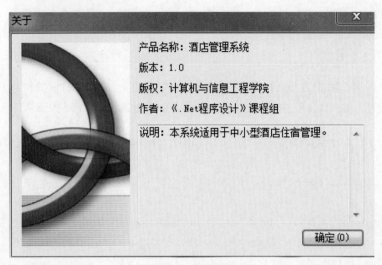

图 8-10 关于界面

2. 功能实现

当单击"关于"窗体的"确定"按钮时,将退出"关于"界面。"确定"按钮事件处理程序代码如下:

```
private void okButton_Click(object sender, EventArgs e)
{
    this.Close();
}
```

第 9 章　实践项目——学生选课管理系统

本章作为课程设计的高级案例。开发学生选课管理系统可以提高学生选课效率，提供更多选课信息，优化选课流程，提升教务管理效率，并为学校提供数据支持决策，从而提高教学质量和学生满意度，促进教育信息化发展。

（1）提高学生选课效率。学生选课管理系统可以提供在线选课功能，使学生能够方便快捷地浏览和选择课程，提高选课效率。

（2）提供更多选课信息。学生选课管理系统可以提供详细的课程信息，包括课程名称、教师信息、上课时间地点、课程描述等。学生可以根据自己的兴趣和需求，更全面地了解课程内容，做出更好的选课决策。

（3）优化选课流程。学生选课管理系统可以提供智能化的选课规划和冲突检测功能，帮助学生避免选课时间冲突或者选课冲突。

（4）提升教务管理效率。学生选课管理系统可以自动化处理选课数据，减少人工操作和纸质文件的使用。教务管理人员可以更方便地管理和统计选课数据，提高工作效率。

（5）提供支持决策。学生选课管理系统可以收集和分析选课数据，为学校提供决策支持。通过分析学生选课偏好和课程需求，学校可以优化课程设置和资源分配，提高教学质量和学生满意度。

9.1　需求分析

学生选课管理系统是一个 C/S 架构的软件项目，分为前台选课和后台管理两部分。

1. 前台选课

前台选课用户是学生，主要具有登录、维护个人密码、选课、查看课程表等功能。具体功能描述如下所述。

（1）登录。学生用户在登录时，需要选择用户身份为学生，然后输入学号和密码进行验证。登录成功后，进入学生选课管理系统前台主窗体；若登录不成功，将显示提示信息。

（2）维护个人密码。学生可以修改个人密码。输入新密码和确认新密码，若两次密码输入一致，实现密码修改；若两次密码输入不一致，则显示提示信息。若用户单击"取消"按钮时，将关闭当前窗体。

（3）选课。当选课时，学生可以通过开课学期查询本专业的全部课程信息，包括课程号和课程名等。当在表格中选择某门课程后，可以查看本门课程的全部详细信息。学生在表格中选中某门课程后，单击"确认"按钮，将提交选课信息。

（4）查看课程表。学生可以查看课程表，包括课程号、课程名、授课教师和课程教室等详细信息。

2. 后台管理

后台管理用户是管理员，主要具有登录、系统信息管理、添加学生信息、修改学生信息、删除学生信息、添加课程信息、修改课程信息、删除课程信息、维护成绩信息等功能。具体功能描述如下所述。

（1）登录。管理员用户在登录时，需要选择用户身份为管理员，然后输入工号和密码进行验证。登录成功后，进入学生选课管理系统后台主窗体；若登录不成功，将显示提示信息。

（2）系统信息管理。管理员可以修改个人密码。输入新密码和确认新密码，单击"修改"按钮，若两次密码输入一致，则实现管理员密码修改；若两次密码输入不一致，则显示提示信息。若用户单击"退出"按钮时，将关闭当前窗体。

（3）添加学生信息。管理员在文本框中输入学生的姓名、学号，然后选择学院、专业、年级，单击"添加"按钮，将提交新增学生信息，并自动分配给学生默认初始密码"123456"；若管理员单击"退出"按钮，将关闭当前窗体。

（4）修改学生信息。管理员通过学号查询到学生信息，可以对学生的姓名、学院、专业和年级进行修改。

（5）删除学生信息。管理员通过学号查询到学生信息，单击"删除"按钮时，则显示提示信息"是否确认删除该学生信息？"，若管理员单击"确定"按钮，则实现学生信息删除；若管理员单击"取消"按钮，则将关闭当前窗体。

（6）添加课程信息。管理员在文本框中输入课程的课程名、课程号、学分、课程总人数等信息，并选择上课时间、专业，单击"提交"，将提交新增课程信息；若管理员单击"退出"按钮，则将关闭当前窗体。

（7）修改课程信息。管理员在维护课程信息页面中可以查看到全部课程信息，单击某课程，可以对课程的学分、授课教师、课程学期等信息进行修改。

（8）删除课程信息。管理员在维护课程信息页面中可以查看到全部课程信息，单击某门课程，然后单击"删除"按钮，则显示提示信息"是否确认删除该课程信息？"，若管理员单击"确定"按钮，则实现课程信息删除；若管理员单击"取消"按钮，则将关闭当前窗体。

（9）维护成绩信息。管理员选择学期和课程名，输入学生学号，可以查询到对应的学生成绩信息，可以在文本框中输入成绩，单击"提交"按钮，实现录入或修改学生成绩。

9.2 概要设计

9.2.1 架构设计

学生选课管理系统采用三层架构设计，分为数据访问层、业务逻辑层以及表示层。三层架构中各层之间的结构关系如图 9-1 所示。

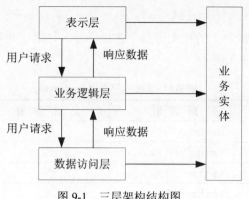

图 9-1 三层架构结构图

9.2.2 功能模块设计

根据需求分析，学生选课管理系统包括两类用户：学生用户和管理员用户。学生用户操作系统前台进行选课，包括登录、个人中心、选课中心和课程中心子模块；管理员用户操作系统后台进行信息管理，包括登录、系统信息管理、学生信息管理、课程信息管理和成绩信息管理子模块。

学生选课管理系统功能模块设计如图 9-2 所示。

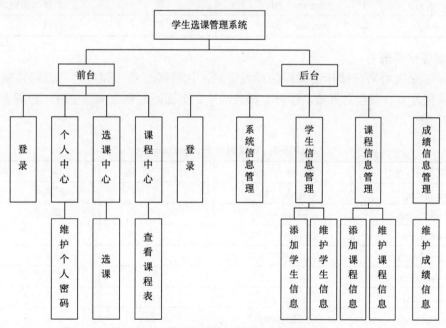

图 9-2 学生选课管理系统功能模块图

9.3 数据库设计

学生选课管理系统采用 MySQL 数据库，名称为 studentcourse，其中共 4 个数据表，分别是学生信息表、管理员信息表、课程信息表、成绩信息表。

1. 学生信息表

学生信息表用来存储学生的相关信息,包括学号、姓名、密码、学院、专业和年级等字段,数据表结构如表 9-1 所示。

表 9-1 学生信息表(students)

字 段 名	数 据 类 型	允 许 空 值	字 段 说 明
stuNum	varchar(30)	否	学号(主键)
stuName	varchar(30)	否	姓名
stuPwd	varchar(30)	否	密码
stuDepart	varchar(50)	否	学院
stuMajor	varchar(50)	否	专业
stuGrade	varchar(30)	否	年级

2. 管理员信息表

管理员信息表用来存储管理员的账号信息,包括工号和密码等字段,数据表结构如表 9-2 所示。

表 9-2 管理员信息表(managers)

字 段 名	数 据 类 型	允 许 空 值	字 段 说 明
mgrNum	varchar(20)	否	工号(主键)
mgrPwd	varchar(20)	否	密码

3. 课程信息表

课程信息表用来存储开设课程的详细信息,包括课程号、课程名、课程学期、授课教师、课程教室、课程总人数、学分、课程周次、上课时间和专业等字段,数据表结构如表 9-3 所示。

表 9-3 课程信息表(courses)

字 段 名	数 据 类 型	允 许 空 值	字 段 说 明
courseNum	varchar(30)	否	课程号(主键)
courseName	varchar(30)	否	课程名
courseTerm	varchar(30)	否	课程学期
courseTeacher	varchar(15)	否	授课教师
courseRoom	varchar(30)	否	课程教室
coursePerson	varchar(10)	否	课程总人数
courseCredit	varchar(9)	否	学分
courseWeeks	varchar(30)	否	课程周次
courseTime	varchar(30)	否	上课时间
courseMajor	varchar(30)	否	专业

4. 成绩信息表

成绩信息表用来存储学生所选课程的成绩信息,包括选课号、学号、学期、课程号、课程名和成绩等字段，数据表结构如表 9-4 所示。

表 9-4　成绩信息表（scores）

字　段　名	数　据　类　型	允　许　空　值	字　段　说　明
selNum	int	否	选课号（主键）
stuNum	varchar（30）	否	学号
courseTerm	varchar（30）	否	学期
courseNum	varchar（30）	否	课程号
courseName	varchar（30）	否	课程名
courseScore	int	是	成绩

9.4　实体模型设计

实体模型类库 Model 由 4 个核心类组成，分别为 Student 类、Manager 类、Course 类和 Score 类。

1. Student 类

Student 类用于封装数据库中的 students 数据表的用户数据，程序代码如下：

```
namespace Model
{
    public class Student
    {
        private string _stuNum;
        private string _stuName;
        private string _stuPwd;
        private string _stuDepart;
        private string _stuMajor;
        private string _stuGrade;
        public string StuNum
        {
            get { return _stuNum; }
            set { _stuNum = value; }
        }
        public string StuName
        {
            get { return _stuName; }
            set { _stuName = value; }
        }
        public string StuPwd
        {
            get { return _stuPwd; }
            set { _stuPwd = value; }
        }
        public string StuDepart
        {
```

```csharp
            get { return _stuDepart; }
            set { _stuDepart = value; }
        }
        public string StuMajor
        {
            get { return _stuMajor; }
            set { _stuMajor = value; }
        }
        public string StuGrade
        {
            get { return _stuGrade; }
            set { _stuGrade = value; }
        }
        public Student()
        {
            _stuNum = "";
            _stuName = "";
            _stuPwd = "";
            _stuDepart = "";
            _stuMajor = "";
            _stuGrade = "";
        }
        public Student(string stuNum, string stuName, string stuPwd, string stuDepart, string stuMajor, string stuGrade)
        {
            _stuNum = stuNum;
            _stuName = stuName;
            _stuPwd = stuPwd;
            _stuDepart = stuDepart;
            _stuMajor = stuMajor;
            _stuGrade = stuGrade;
        }
    }
}
```

2. Manager 类

Manager 类用于封装数据库中的 managers 数据表的用户数据，程序代码如下：

```csharp
namespace Model
{
    public class Manager
    {
        private string _mgrNum;
        private string _mgrPwd;
        public string MgrNum
        {
            get { return _mgrNum; }
            set { _mgrNum = value; }
        }
```

```csharp
        public string MgrPwd
        {
            get { return _mgrPwd; }
            set { _mgrPwd = value; }
        }
        public Manager()
        {
            _mgrNum = "";
            _mgrPwd = "";
        }
        public Manager(string mgrNum, string mgrPwd)
        {
            _mgrNum = mgrNum;
            _mgrPwd = mgrPwd;
        }
    }
}
```

3. Course 类

Course 类用于封装数据库中的 courses 数据表的用户数据，程序代码如下：

```csharp
namespace Model
{
    public class Course
    {
        private string _courseNum;
        private string _courseName;
        private string _courseTerm;
        private string _courseTeacher;
        private string _courseRoom;
        private string _coursePerson;
        private string _courseCredit;
        private string _courseWeeks;
        private string _courseTime;
        private string _courseMajor;
        public string CourseNum
        {
            get { return _courseNum; }
            set { _courseNum = value; }
        }
        public string CourseName
        {
            get { return _courseName; }
            set { _courseName = value; }
        }
        public string CourseTerm
        {
            get { return _courseTerm;
```

```csharp
            set { _courseTerm = value; }
        }
        public string CourseTeacher
        {
            get { return _courseTeacher; }
            set { _courseTeacher = value; }
        }
        public string CourseRoom
        {
            get { return _courseRoom; }
            set { _courseRoom = value; }
        }
        public string CoursePerson
        {
            get { return _coursePerson; }
            set { _coursePerson = value; }
        }
        public string CourseCredit
        {
            get { return _courseCredit; }
            set { _courseCredit = value; }
        }
        public string CourseWeeks
        {
            get { return _courseWeeks; }
            set { _courseWeeks = value; }
        }
        public string CourseTime
        {
            get { return _courseTime; }
            set { _courseTime = value; }
        }
        public string CourseMajor
        {
            get { return _courseMajor; }
            set { _courseMajor = value; }
        }
        public Course()
        {
        }
        public Course(string courseNum, string courseName, string courseTerm,
string courseTeacher, string courseRoom, string coursePerson, string courseCredit,
string courseWeeks, string courseTime, string courseMajor)
        {
            _courseNum = courseNum;
            _courseName = courseName;
            _courseTerm = courseTerm;
```

```
            _courseTeacher = courseTeacher;
            _courseRoom = courseRoom;
            _coursePerson = coursePerson;
            _courseCredit = courseCredit;
            _courseWeeks = courseWeeks;
            _courseTime = courseTime;
            _courseMajor = courseMajor;
        }
    }
}
```

4. Score 类

Score 类用于封装数据库中的 scores 数据表的用户数据，程序代码如下：

```
namespace Model
{
    public class Score
    {
        private int _selNum;
        private string _stuNum;
        private string _courseTerm;
        private string _courseNum;
        private string _courseName;
        private int _courseScore;
        public int SelNum
        {
            get { return _selNum; }
            set { _selNum = value; }
        }
        public string StuNum
        {
            get { return _stuNum; }
            set { _stuNum = value; }
        }
        public string CourseTerm
        {
            get { return _courseTerm; }
            set { _courseTerm = value; }
        }
        public string CourseNum
        {
            get { return _courseNum; }
            set { _courseNum = value; }
        }
        public string CourseName
        {
            get { return _courseName; }
            set { _courseName = value; }
```

```
        }
        public int CourseScore
        {
            get { return _courseScore; }
            set { _courseScore = value; }
        }
    }
}
```

9.5 数据访问层设计

数据访问层类库 DAL 由一个核心类 DBOper.cs 组成，其中封装了执行数据的增、删、改、查等功能的方法。

1. 配置 APP.Config

```
<configuration>
  <connectionStrings>
    <add name = "conStr" connectionString = "server = localhost;user id = root;password = 123456;database = studentcourse;Charset = utf8;" />
  </connectionStrings>
</configuration>
```

2. DBOper 类

```
using MySql.Data.MySqlClient;
using System.Configuration;
using System.Data;
namespace DAL
{
    public class DBOper
    {
        private static MySqlConnection conn;
        public static MySqlConnection Connection
        {
            get
            {
                if (conn == null)
                {
                    string connStr = ConfigurationManager.ConnectionStrings["conStr"].ConnectionString;
                    conn = new MySqlConnection(connStr);
                    conn.Open();
                }
                else if (conn.State == ConnectionState.Closed)
                {
                    conn.Open();
                }
                else if (conn.State == ConnectionState.Broken)
```

```csharp
            {
                conn.Close();
                conn.Open();
            }
            return conn;
        }
    }
    public static int ExecuteCommand(string sql)
    {
        MySqlCommand cmd = new MySqlCommand(sql, Connection);
        return cmd.ExecuteNonQuery();
    }
    public static int ExecuteCommand(string sql, params MySqlParameter[] values)
    {
        MySqlCommand cmd = new MySqlCommand(sql, Connection);
        cmd.Parameters.AddRange(values);
        return cmd.ExecuteNonQuery();
    }
    public static MySqlDataReader GetReader(string sql)
    {
        MySqlCommand cmd = new MySqlCommand(sql, Connection);
        return cmd.ExecuteReader();
    }
    public static MySqlDataReader GetReader(string sql, params MySqlParameter[] values)
    {
        MySqlCommand cmd = new MySqlCommand(sql, Connection);
        cmd.Parameters.AddRange(values);
        return cmd.ExecuteReader();
    }
    public static DataTable GetDataTable(string sql,string table)
    {
        MySqlDataAdapter da = new MySqlDataAdapter(sql, Connection);
        DataSet ds = new DataSet();
        da.Fill(ds,table);
        return ds.Tables[table];
    }
    public static DataTable GetDataTable(string sql, params MySqlParameter[] values)
    {
        MySqlCommand cmd = new MySqlCommand(sql, Connection);
        cmd.Parameters.AddRange(values);
        MySqlDataAdapter da = new MySqlDataAdapter(cmd);
        DataSet ds = new DataSet();
        da.Fill(ds);
```

```
            return ds.Tables[0];
        }
    }
}
```

9.6 业务逻辑层设计

业务逻辑层类库 BLL 由 5 个核心类组成，分别为 RoleManager 类、ManagerManager 类、StudentManager 类、CourseManager 类和 ScoreManager 类。RoleManager 类用于保存当前登录用户信息；ManagerManager 类用于对管理员进行管理；StudentManager 类用于对学生进行管理；CourseManager 类用于对课程进行管理；ScoreManager 类用于对成绩进行管理。

1. RoleManager 类

```
using Model;
namespace BLL
{
    public class RoleManager
    {
        //登录后保存登录用户
        public static Student curStu = new Student();
        public static Manager curMgr = new Manager();
    }
}
```

2. ManagerManager 类

```
using System;
using Model;
using MySql.Data.MySqlClient;
using DAL;
namespace BLL
{
    public class ManagerManager
    {
        //1.管理员登录时，根据工号和密码查询管理员
        public static Manager GetManager(string mgrNum, string mgrPwd)
        {
            Manager manager = null;
            string sql = "select * from managers where mgrNum = @mgrNum and mgrPwd = @mgrPwd";
            MySqlParameter p1 = new MySqlParameter("@mgrNum", mgrNum);
            MySqlParameter p2 = new MySqlParameter("@mgrPwd", mgrPwd);
            MySqlDataReader reader = DBOper.GetReader(sql, p1, p2);
            if (reader.Read())
            {
                manager = new Manager();
                manager.MgrNum = Convert.ToString(reader["mgrNum"]);
```

```csharp
                manager.MgrPwd = Convert.ToString(reader["mgrPwd"]);
            }
            reader.Close();
            return manager;
        }
        //2.管理员修改个人密码
        public static bool ModifyPwd(string newPwd, string mgrNum)
        {
            string sql = "update managers set mgrPwd = @mgrPwd where mgrNum = @mgrNum";
            MySqlParameter p1 = new MySqlParameter("@mgrPwd", newPwd);
            MySqlParameter p2 = new MySqlParameter("@mgrNum", mgrNum);
            if (DBOper.ExecuteCommand(sql, p1, p2) == 1)
            {
                return true;
            }
            else
            {
                return false;
            }
        }
    }
}
```

3. StudentManager 类

```csharp
using System;
using Model;
using MySql.Data.MySqlClient;
using DAL;
using System.Data;
namespace BLL
{
    public static class StudentManager
    {
        //1.学生登录时，根据学号与密码查询学生用户
        public static Student GetStudent(string stuNum, string stuPwd)
        {
            Student stu = null;
            string sql = "select * from students where stuNum = @stuNum and stuPwd = @stuPwd";
            MySqlParameter p1 = new MySqlParameter("@stuNum", stuNum);
            MySqlParameter p2 = new MySqlParameter("@stuPwd", stuPwd);
            MySqlDataReader reader = DBOper.GetReader(sql, p1, p2);
            if (reader.Read())
            {
                stu = new Student();
                stu.StuNum = Convert.ToString(reader["stuNum"]);
```

```csharp
            stu.StuPwd = Convert.ToString(reader["stuPwd"]);
            stu.StuName = Convert.ToString(reader["stuName"]);
            stu.StuDepart = Convert.ToString(reader["stuDepart"]);
            stu.StuMajor = Convert.ToString(reader["stuMajor"]);
            stu.StuGrade = Convert.ToString(reader["stuGrade"]);
        }
        reader.Close();
        return stu;
    }
    //2.学生用户修改个人密码
    public static bool UpdateStuPassword(string stuNum, string stuPwd)
    {
        string sqlUpdateUser = @"update students set stuPwd = @stuPwd where stuNum = @stuNum";
        MySqlParameter p1 = new MySqlParameter("@stuPwd", stuPwd);
        MySqlParameter p2 = new MySqlParameter("@stuNum", stuNum);
        MySqlParameter[] paramArray = new MySqlParameter[] { p1, p2 };
        if (DBOper.ExecuteCommand(sqlUpdateUser, paramArray) == 1)
        {
            return true;
        }
        else
        {
            return false;
        }
    }
    //3.管理员添加学生信息
    public static bool AddStudent(Student stu)
    {
        string sql = "insert into students values (@stuNum,@stuName,@stuPwd,@stuDepart,@stuMajor,@stuGrade)";
        MySqlParameter p1 = new MySqlParameter("@stuNum", stu.StuNum);
        MySqlParameter p2 = new MySqlParameter("@stuName", stu.StuName);
        MySqlParameter p3 = new MySqlParameter("@stuPwd", stu.StuPwd);
        MySqlParameter p4 = new MySqlParameter("@stuDepart", stu.StuDepart);
        MySqlParameter p5 = new MySqlParameter("@stuMajor", stu.StuMajor);
        MySqlParameter p6 = new MySqlParameter("@stuGrade", stu.StuGrade);
        MySqlParameter[] p = new MySqlParameter[] { p1, p2, p3, p4, p5, p6 };
        if (DBOper.ExecuteCommand(sql, p) == 1)
        {
            return true;
        }
        else
        {
            return false;
        }
```

```csharp
        }
        //4.根据学号查询学生信息
        public static DataTable GetStudent(string stuNum)
        {
            string sql = "select * from students where stuNum = @stuNum";
            MySqlParameter p = new MySqlParameter("@stuNum", stuNum);
            DataTable dataTable = DBOper.GetDataTable(sql, p);
            return dataTable;
        }
        //5.管理员修改学生信息
        public static bool UpdateStudent(Student stu)
        {
            string sqlUpate = "update students set stuName=@stuName,stuDepart = @stuDepart,stuMajor = @stuMajor,stuGrade = @stuGrade where stuNum = @stuNum ";
            MySqlParameter p4 = new MySqlParameter("@stuName", stu.StuName);
            MySqlParameter p3 = new MySqlParameter("@stuDepart", stu.StuDepart);
            MySqlParameter p5 = new MySqlParameter("@stuMajor", stu.StuMajor);
            MySqlParameter p2 = new MySqlParameter("@stuGrade", stu.StuGrade);
            MySqlParameter p6 = new MySqlParameter("@stuNum", stu.StuNum);
            MySqlParameter[] p = { p2, p3, p4, p5, p6 };
            if (DBOper.ExecuteCommand(sqlUpate, p) == 1)
            {
                return true;
            }
            else
            {
                return false;
            }
        }
        //6.管理员删除学生信息
        public static bool DeleteStudent(Student stu)
        {
            string sql = "delete from students where stuNum = @stuNum";
            MySqlParameter p = new MySqlParameter("@stuNum", stu.StuNum);
            if (DBOper.ExecuteCommand(sql, p) == 1)
            {
                return true;
            }
            else
            {
                return false;
            }
        }
    }
}
```

4. CourseManager 类

```csharp
using Model;
using MySql.Data.MySqlClient;
using System;
using System.Collections.Generic;
using DAL;
using System.Data;
namespace BLL
{
    public class CourseManager
    {
        //1.根据课程名查询课程信息
        public static DataTable SearchCourse(string courseNum)
        {
            string sql = "select * from courses where courseNum = @courseNum";
            MySqlParameter p = new MySqlParameter("courseNum", courseNum);
            DataTable dt = DBOper.GetDataTable(sql, p);
            return dt;
        }
        //2.管理员添加课程信息
        public static bool AddCourse(Course course)
        {
            string sql = "insert into courses values (@courseNum, @courseName, @courseTerm, @courseTeacher, @courseRoom, @coursePerson, @courseCredit, @courseWeeks, @courseTime, @courseMajor)";
            MySqlParameter p1 = new MySqlParameter("@courseNum", course.CourseNum);
            MySqlParameter p2 = new MySqlParameter("@courseName", course.CourseName);
            MySqlParameter p3 = new MySqlParameter("@courseTerm", course.CourseTerm);
            MySqlParameter p4 = new MySqlParameter("@courseTeacher", course.CourseTeacher);
            MySqlParameter p5 = new MySqlParameter("@courseRoom", course.CourseRoom);
            MySqlParameter p6 = new MySqlParameter("@coursePerson", course.CoursePerson);
            MySqlParameter p7 = new MySqlParameter("@courseCredit", course.CourseCredit);
            MySqlParameter p8 = new MySqlParameter("@courseWeeks", course.CourseWeeks);
            MySqlParameter p9 = new MySqlParameter("@courseTime", course.CourseTime);
            MySqlParameter p10 = new MySqlParameter("@courseMajor", course.CourseMajor);
```

```csharp
            if (DBOper.ExecuteCommand(sql, p1, p2, p3, p4, p5, p6, p7, p8, p9, p10) == 1)
            {
                return true;
            }
            else
            {
                return false;
            }
        }
        //3.获取全部课程信息
        public static DataTable GetCourseInfo()
        {
            string sql = @"select courseNum as 课程号,courseName as 课程名,courseTerm as 课程学期,courseTeacher as 授课教师,courseRoom as 上课教室,coursePerson as 课程总人数,courseCredit as 学分,courseWeeks as 课程周次,courseTime as 上课时间,courseMajor as 专业 from courses";
            DataTable dt = DBOper.GetDataTable(sql, "courses");
            return dt;
        }
        //4.管理员修改课程信息
        public static bool UpdateCourse(Course course)
        {
            string sqlUpate = "update courses set courseName = @courseName,courseTerm = @courseTerm,courseTeacher = @courseTeacher,courseRoom = @courseRoom,coursePerson = @coursePerson,courseCredit = @courseCredit,courseWeeks = @courseWeeks,courseTime = @courseTime,courseMajor = @courseMajor where courseNum = @courseNum";
            MySqlParameter p1 = new MySqlParameter("@courseNum", course.CourseNum);
            MySqlParameter p2 = new MySqlParameter("@courseName", course.CourseName);
            MySqlParameter p3 = new MySqlParameter("@courseTerm", course.CourseTerm);
            MySqlParameter p4 = new MySqlParameter("@courseTeacher", course.CourseTeacher);
            MySqlParameter p5 = new MySqlParameter("@courseRoom", course.CourseRoom);
            MySqlParameter p6 = new MySqlParameter("@coursePerson", course.CoursePerson);
            MySqlParameter p7 = new MySqlParameter("@courseCredit", course.CourseCredit);
            MySqlParameter p8 = new MySqlParameter("@courseWeeks", course.CourseWeeks);
            MySqlParameter p9 = new MySqlParameter("@courseTime", course.CourseTime);
```

```csharp
            MySqlParameter p10 = new MySqlParameter("@courseMajor", 
course.CourseMajor);
            MySqlParameter[] p = { p1, p2, p3, p4, p5, p6, p7, p8, p9, p10 };
            if (DBOper.ExecuteCommand(sqlUpate, p) == 1)
            {
                return true;
            }
            else
            {
                return false;
            }
        }
        //5.管理员删除课程信息
        public static bool DeleteCourse(Course course)
        {
            string sqlDelete = "delete from courses where courseName = @courseName and courseNum = @courseNum";
            MySqlParameter p1 = new MySqlParameter("@courseName", 
course.CourseName);
            MySqlParameter p2 = new MySqlParameter("@courseNum", 
course.CourseNum);
            if (DBOper.ExecuteCommand(sqlDelete, p1, p2) == 1)
            {
                return true;
            }
            else
            {
                return false;
            }
        }
        //6.学生选课时，根据专业和学期，查询课程信息
        public static DataTable GetAllCourseInfo(string courseTerm, string courseMajor)
        {
            string sql = @"select courseNum as 课程号,courseName as 课程名,courseTerm as 课程学期,courseTeacher as 授课教师,courseMajor as 专业 from courses where courseTerm like '%" + courseTerm + "%' and courseMajor like '%" + courseMajor + "%'";
            DataTable dt = DBOper.GetDataTable(sql);
            return dt;
        }
        //7.学生选课时，根据学生选择的课程号查询课程详细信息
        public static DataTable GetFullCourseInfo(string courseNum)
        {
            string sql = @"select courseName as 课程名,courseTime as 上课时间,courseRoom as 课程教室,coursePerson as 上课总人数,courseCredit as 学分,courseWeeks
```

```csharp
            as 课程周次,courseMajor as 专业 from courses where courseNum = @courseNum";
            MySqlParameter param = new MySqlParameter("@CourseNum", courseNum);
            DataTable dt = DBOper.GetDataTable(sql, param);
            return dt;
        }
        //8.学生查看课程表时，根据学号获得该学生所有选择课程
        public static List<string> GetCourseByStuNum(string stuNum)
        {
            List<string> courseNumList = new List<string>();
            string sql = "select courseNum from scores where stuNum = @stuNum";
            MySqlParameter[] parameters = { new MySqlParameter("@StuNum", stuNum) };
            using (MySqlDataReader reader = DBOper.GetReader(sql, parameters))
            {
                while (reader.Read())
                {
                    string courseNum = Convert.ToString(reader["courseNum"]);
                    courseNumList.Add(courseNum);
                }
            }
            return courseNumList;
        }
        //9.学生查看课表时，根据课程号获得全部课程信息
        public static DataTable GetCourseDetails(string courseNum)
        {
            string sql = "select courseNum,courseName,courseTerm,courseTeacher,courseRoom,coursePerson,courseCredit,courseWeeks,courseTime,courseMajor from courses where courseNum = @CourseNum";
            MySqlParameter param = new MySqlParameter("@CourseNum", courseNum);
            DataTable dt = DBOper.GetDataTable(sql, param);
            return dt;
        }
    }
}
```

5. ScoreManager

```csharp
using DAL;
using Model;
using MySql.Data.MySqlClient;
using System.Data;
namespace BLL
{
    public class ScoreManager
    {
        //1.学生选课时，填加到成绩信息表即选择该课程，其中成绩值为空
        public static bool AddSelectCourse(Score score)
```

```csharp
        {
            string sql = "insert into scores values (@selNum,@stuNum,
@courseTerm,@courseNum,@courseName,@curseScore)";
            MySqlParameter p1 = new MySqlParameter("@selNum", null);
            MySqlParameter p2 = new MySqlParameter("@stuNum", score.StuNum);
            MySqlParameter p3 = new MySqlParameter("@courseTerm",
score.CourseTerm);
            MySqlParameter p4 = new MySqlParameter("@courseNum",
score.CourseNum);
            MySqlParameter p5 = new MySqlParameter("@courseName",
score.CourseName);
            MySqlParameter p6 = new MySqlParameter("@curseScore", null);
            if (DBOper.ExecuteCommand(sql, p1, p2, p3, p4, p5, p6) == 1)
            {
                return true;
            }
            else
            {
                return false;
            }
        }
        //2.管理员的维护成绩窗体加载时，填充课程学期和课程名到窗体的下拉列表框
        public static MySqlDataReader GetTermAndName()
        {
            string sql = "select distinct courseName,courseTerm from scores";
            return DBOper.GetReader(sql);
        }
        //3.管理员维护成绩时，根据用户输入查询课程信息
        public static DataTable GetScoreInfo(string stuNum, string courseName,
string courseTerm)
        {
            string sql = "select stuNum as 学号,courseTerm as 学期,courseNum
as 课程号,courseName as 课程名,courseScore as 成绩 from scores where stuNum =
@stuNum and courseName = @courseName and courseTerm = @courseTerm";
            MySqlParameter p1 = new MySqlParameter("@stuNum", stuNum);
            MySqlParameter p2 = new MySqlParameter("@courseName", courseName);
            MySqlParameter p3 = new MySqlParameter("@courseTerm", courseTerm);
            MySqlParameter[] p = { p1, p2, p3 };
            DataTable dt = DBOper.GetDataTable(sql, p);
            return dt;
        }
        //4.管理员维护成绩时，提交或修改成绩
        public static bool UpdateScore(string courseName, string stuNum, int
courseScore)
        {
```

```
                string sql = "update scores set courseScore = @courseScore where 
courseName = @courseName and stuNum = @stuNum";
                MySqlParameter p1 = new MySqlParameter("@courseName", courseName);
                MySqlParameter p2 = new MySqlParameter("@stuNum", stuNum);
                MySqlParameter p3 = new MySqlParameter("@courseScore", courseScore);
                MySqlParameter[] p = { p1, p2, p3 };
                if (DBOper.ExecuteCommand(sql, p) == 1)
                {
                    return true;
                }
                else
                {
                    return false;
                }
            }
        }
    }
```

9.7 表示层设计

学生选课管理系统中的模块和对应的窗体名称如表 9-5 所示。

表 9-5 学生选课管理系统窗体描述表

模 块		窗体 Text 属性值	窗体 Name 属性值
登录		登录	LoginForm
前台	系统前台主窗体	学生选课管理系统——前台	StudentMainForm
	个人中心	维护个人密码	MaintainPassword
	选课中心	选课	SelectCourse
	课程中心	查看课程表	ShowCourseSchedule
后台	系统后台主窗体	学生选课管理系统——后台	ManagerMainForm
	系统信息管理	系统信息管理	ManageSystemInfo
	学生信息管理	添加学生信息	AddStudentinfo
		维护学生信息	MaintainStudentinfo
	课程信息管理	添加课程信息	AddCourseInfo
		维护课程信息	MaintainCourseInfo
	成绩信息管理	维护成绩信息	MaintainScoreInfo

9.7.1 登录设计

1. 窗体设计

当用户选择学生身份时,"登录"窗体如图 9-3 所示。当用户选择管理员身份时,"登录"窗体如图 9-4 所示。

图 9-3 学生用户的"登录"窗体

图 9-4 管理员用户的"登录"窗体

"登录"窗体和主要控件的属性设置如表 9-6 所示。

表 9-6 "登录"窗体和主要控件的属性设置

Name	类型	Text 属性	其他属性
LoginForm	Form	登录	ControlBox：True StartPosition：CenterScreen
pictureBox1	PictureBox		Image 设置为选定的图片 SizeMode：StretchImage
groupBox1	GroupBox	登录	
stuRgb	RadioButton	学生	Checked：True
mgrRgb	RadioButton	管理员	
numTxt	TextBox		
pwdtxt	TextBox		UseSystemPasswordChar：True
loginBtn	Button	登录	
exitBtn	Button	退出	

2. 功能实现代码

```csharp
using System;
using System.Windows.Forms;
using BLL;
using Model;
namespace StudentCourseWindowsFormsApp
{
    public partial class LoginForm : Form
    {
        public LoginForm()
        {
            InitializeComponent();
        }
        private void loginBtn_Click(object sender, EventArgs e)
        {
            string userName = numTxt.Text;
            string password = pwdTxt.Text;
            if (string.IsNullOrEmpty(userName) || string.IsNullOrEmpty(password))
            {
                MessageBox.Show("用户名或密码不能为空!", "提示", MessageBoxButtons.OK, MessageBoxIcon.Information);
                return;
            }
            if (stuRgb.Checked == true)
            {
                Student stu = StudentManager.GetStudent(userName, password);
                if (stu == null)
                {
                    MessageBox.Show("用户名或密码错误!", "提示", MessageBoxButtons.OK, MessageBoxIcon.Information);
                }
                else
                {
                    RoleManager.curStu = stu;
                    this.Hide();
                    StudentMainForm form = new StudentMainForm();
                    form.Show();
                }
            }
            else if (mgrRgb.Checked == true)
            {
                Manager manager = ManagerManager.GetManager(userName, password);
                if (manager == null)
                {
```

```csharp
                    MessageBox.Show("用户名或密码错误!", "提示", 
MessageBoxButtons.OK, MessageBoxIcon.Information);
                }
                else
                {
                    RoleManager.curMgr = manager;
                    this.Hide();
                    ManagerMainForm mainForm = new ManagerMainForm();
                    mainForm.Show();
                }
            }
            else
            {
                MessageBox.Show("请选择您的登录身份!");
                return;
            }
        }
        private void exitBtn_Click(object sender, EventArgs e)
        {
            //退出系统
            this.Close();
        }
        private void LoginForm_Load(object sender, EventArgs e)
        {
            label1.Text = "学号: ";
        }
        private void mgrRgb_CheckedChanged(object sender, EventArgs e)
        {
            if (mgrRgb.Checked)
            {
                label1.Text = "工号: ";
            }
            else
            {
                label1.Text = "学号: ";
            }
        }
    }
}
```

9.7.2 系统前台主窗体设计

1. 窗体设计

"学生选课管理系统——前台"主窗体如图9-5所示。

图 9-5 "学生选课管理系统——前台"主窗体

"学生选课管理系统——前台"主窗体和主要控件的属性设置如表 9-7 所示。

表 9-7 "学生选课管理系统——前台"主窗体和主要控件的属性设置

Name	类　型	其 他 属 性
StudentMainForm	Form	WindowState：Maximized IsMdiContainer：True
toolStripMenuItem1	ToolStripMenuItem	Text：个人中心
MaintainPasswordToolStripMenuItem	ToolStripMenuItem	Text：维护个人密码
toolStripMenuItem2	ToolStripMenuItem	Text：选课中心
SelectCourseToolStripMenuItem	ToolStripMenuItem	Text：选课
toolStripMenuItem3	ToolStripMenuItem	Text：课程中心
ShowCourseScheduleToolMenuItem	ToolStripMenuItem	Text：查看课程表

2. 功能实现代码

```
using System;
using System.Windows.Forms;
namespace StudentCourseWindowsFormsApp
{
    public partial class StudentMainForm : Form
    {
        public StudentMainForm()
        {
            InitializeComponent();
        }
        private void MaintainPasswordToolStripMenuItem_Click(object sender, EventArgs e)
```

```
        {
            MaintainPassword form = new MaintainPassword();
            form.MdiParent = this;
            form.Show();
        }
        private void SelectCourseToolStripMenuItem_Click(object sender, EventArgs e)
        {
            SelectCourse form = new SelectCourse();
            form.MdiParent = this;
            form.Show();
        }
        private void ShowCourseScheduleToolStripMenuItem_Click(object sender, EventArgs e)
        {
            ShowCourseSchedule form = new ShowCourseSchedule();
            form.MdiParent = this;
            form.Show();
        }
    }
}
```

9.7.3 维护个人密码设计

1. 窗体设计

"维护个人密码"窗体如图 9-6 所示。

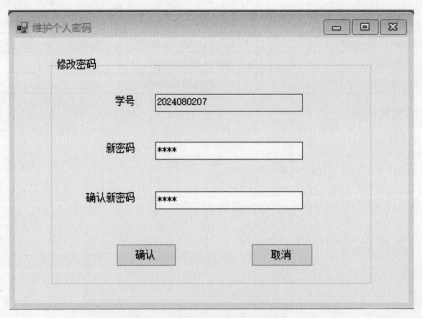

图 9-6 "维护个人密码"窗体

"维护个人密码"窗体和主要控件的属性设置如表9-8所示。

表9-8 "维护个人密码"窗体和主要控件的属性设置

Name	类 型	Text 属性	其 他 属 性
MaintainPassword	Form	维护个人密码	
groupBox1	GroupBox	修改密码	
stuNumTxt	TextBox		ReadOnly：True
stuPwdTxt	TextBox		UseSystemPasswordChar：True
stuPwdSureTxt	TextBox		UseSystemPasswordChar：True
sureBtn	Button	确认	
cancleBtn	Button	取消	

2. 功能实现代码

```csharp
using System;
using System.Windows.Forms;
using BLL;
namespace StudentCourseWindowsFormsApp
{
    public partial class MaintainPassword : Form
    {
        public MaintainPassword()
        {
            InitializeComponent();
        }
        private void sureBtn_Click(object sender, EventArgs e)
        {
            if (stuNumTxt.Text.Trim() == "" || stuPwdTxt.Text.Trim() == "")
            {
                MessageBox.Show("请将信息填写完整!");
            }
            else
            {
                if (stuPwdTxt.Text.Trim() != stuPwdSureTxt.Text.Trim())
                {
                    MessageBox.Show("两次输入的密码不一致,请重新输入!");
                }
                else
                {
                    string stuNum = stuNumTxt.Text;
                    string stuPwd = stuPwdTxt.Text;
                    if (StudentManager.UpdateStuPassword(stuNum, stuPwd))
                    {
                        MessageBox.Show("修改成功!");
                    }
```

```
                else
                {
                    MessageBox.Show("修改失败！");
                }
            }
        }
        private void cancelBtn_Click(object sender, EventArgs e)
        {
            this.Close();
        }
        private void MaintainPassword_Load(object sender, EventArgs e)
        {
            stuNumTxt.Text = RoleManager.curStu.StuNum;
        }
    }
}
```

9.7.4 选课设计

1. 窗体设计

"选课"窗体如图 9-7 所示。

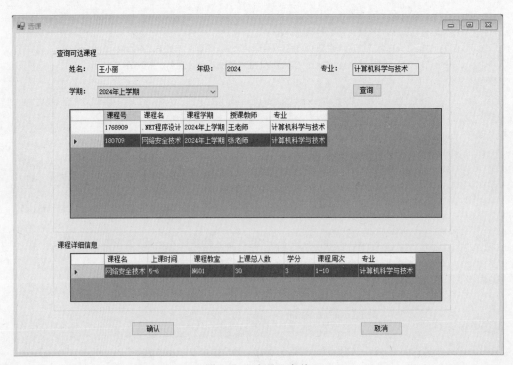

图 9-7 "选课"窗体

"选课"窗体和主要控件的属性设置如表 9-9 所示。

表 9-9 "选课"窗体和主要控件的属性设置

Name	类　型	Text 属性	其　他　属　性
SelectCourse	Form	选课	
groupBox1	GroupBox	查询可选课程	
groupBox2	GroupBox	课程详细信息	
stuNameTxt	TextBox		ReadOnly：True
stuGradetxt	TextBox		ReadOnly：True
stuMajorTxt	TextBox		ReadOnly：True
courseTermCom	ComboBox		DropDownStyle：DropDownList
searchBtn	Button	查询	
searchCourseDgv	DataGridView		
courseDetailDgv	DataGridView		
sureBtn	Button	确认	
cancleBtn	Button	取消	

2. 功能实现代码

```
using System;
using System.Data;
using System.Windows.Forms;
using BLL;
using Model;
namespace StudentCourseWindowsFormsApp
{
    public partial class SelectCourse : Form
    {
        public SelectCourse()
        {
            InitializeComponent();
        }
        private void sureBtn_Click(object sender, EventArgs e)
        {
            string courseNum = searchCourseDgv.SelectedRows[0].Cells["课程号"].Value.ToString();
            string courseName = searchCourseDgv.SelectedRows[0].Cells["课程名"].Value.ToString();
            string courseTerm = searchCourseDgv.SelectedRows[0].Cells["课程学期"].Value.ToString();
            string stuNum = RoleManager.curStu.StuNum;
            Score score = new Score();
            score.StuNum = stuNum;
            score.CourseTerm = courseTerm;
            score.CourseNum = courseNum;
            score.CourseName = courseName;
            if (ScoreManager.AddSelectCourse(score))
```

```csharp
            {
                MessageBox.Show("选课成功！");
            }
            else
            {
                MessageBox.Show("选课失败！");
            }
        }
        private void searchBtn_Click(object sender, EventArgs e)
        {
            string courseTerm = courseTermCom.Text;
            string courseMajor = stuMajorTxt.Text;
            DataTable dt = CourseManager.GetAllCourseInfo(courseTerm, courseMajor);
            searchCourseDgv.DataSource = dt;
        }
        private void SelectCourse_Load(object sender, EventArgs e)
        {
            stuNameTxt.Text = RoleManager.curStu.StuName;
            stuGradeTxt.Text = RoleManager.curStu.StuGrade;
            stuMajorTxt.Text = RoleManager.curStu.StuMajor;
            courseTermCom.Items.Add("2020年上学期");
            courseTermCom.Items.Add("2020年下学期");
            courseTermCom.Items.Add("2021年上学期");
            courseTermCom.Items.Add("2021年下学期");
            courseTermCom.Items.Add("2022年上学期");
            courseTermCom.Items.Add("2022年下学期");
            courseTermCom.Items.Add("2023年上学期");
            courseTermCom.Items.Add("2023年下学期");
            courseTermCom.Items.Add("2024年上学期");
            courseTermCom.Items.Add("2024年下学期");
            courseTermCom.Items.Add("2025年上学期");
            courseTermCom.Items.Add("2026年下学期");
        }
        private void searchCourseDgv_CellContentClick(object sender, DataGridViewCellEventArgs e)
        {
            if (searchCourseDgv.SelectedRows.Count > 0)
            {
                DataGridViewRow selectedRow = searchCourseDgv.SelectedRows[0];
                string courseNum = selectedRow.Cells[0].Value.ToString();
                sureBtn.Enabled = true;
                DataTable dt = CourseManager.GetFullCourseInfo(courseNum);
                courseDetailDgv.DataSource = dt;
            }
```

```
            else
            {
                MessageBox.Show("请先选择一门课程!");
            }
        }
        private void cancelBtn_Click(object sender, EventArgs e)
        {
            this.Hide();
        }
    }
}
```

9.7.5 查看课程表设计

1. 窗体设计

"查看课程表"窗体如图 9-8 所示。

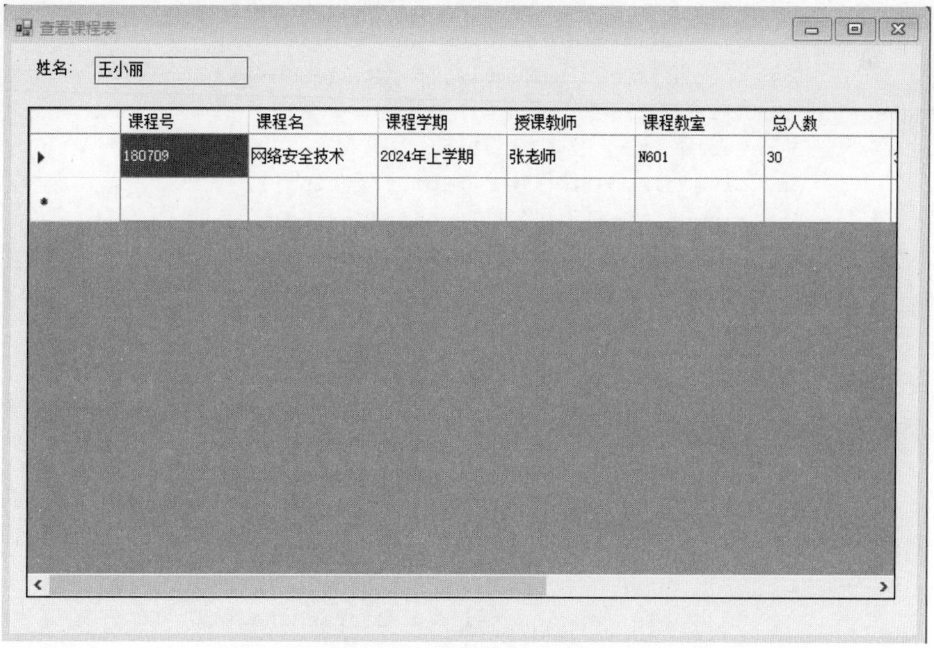

图 9-8 "查看课程表"窗体

"查看课程表"窗体和主要控件的属性设置如表 9-10 所示。

表 9-10 "查看课程表"窗体和主要控件的属性设置

Name	类　型	Text 属性	其 他 属 性
ShowCourseSchedule	Form	查看课程表	
txtName	TextBox		ReadOnly：True
dgvCourse	DataGridView		

2. 功能实现代码

```csharp
using System;
using System.Collections.Generic;
using System.Data;
using System.Windows.Forms;
using BLL;
namespace StudentCourseWindowsFormsApp
{
    public partial class ShowCourseSchedule : Form
    {
        public ShowCourseSchedule()
        {
            InitializeComponent();
        }
        private void ShowCourseSchedule_Load(object sender, EventArgs e)
        {
            txtName.Text = RoleManager.curStu.StuName;
            string stuNum = RoleManager.curStu.StuNum;
            List<string> courseNumList = CourseManager.GetCourseByStuNum(stuNum);
            DataTable dt = new DataTable();
            dt.Columns.Add("课程号", typeof(string));
            dt.Columns.Add("课程名", typeof(string));
            dt.Columns.Add("课程学期", typeof(string));
            dt.Columns.Add("授课教师", typeof(string));
            dt.Columns.Add("课程教室", typeof(string));
            dt.Columns.Add("总人数", typeof(string));
            dt.Columns.Add("学分", typeof(string));
            dt.Columns.Add("课程周次", typeof(string));
            dt.Columns.Add("上课时间", typeof(string));
            dt.Columns.Add("专业", typeof(string));
            foreach (string courseNum in courseNumList)
            {
                DataTable courseDetails = CourseManager.GetCourseDetails(courseNum);
                if (courseDetails != null && courseDetails.Rows.Count > 0)
                {
                    dt.Rows.Add(courseDetails.Rows[0].ItemArray);
                }
            }
            dgvCourse.DataSource = dt;
        }
    }
}
```

9.7.6 系统后台主窗体设计

1. 窗体设计

"学生选课管理系统——后台"主窗体如图9-9所示。

图9-9 "学生选课管理系统——后台"主窗体

"学生选课管理系统——后台"主窗体和主要控件的属性设置如表9-11所示。

表9-11 "学生选课管理系统——后台"主窗体和主要控件的属性设置

Name	类 型	其 他 属 性
ManagerMainForm	Form	WindowState：Maximized IsMdiContainer：True
ManagerSystemInfoToolStripMenuItem	ToolStripMenuItem	Text：系统信息管理
menuStu	ToolStripMenuItem	Text：学生信息管理
AddStudentinfoToolStripMenuItem	ToolStripMenuItem	Text：添加学生信息
MaintainStudentinfoToolStripMenuItem	ToolStripMenuItem	Text：维护学生信息
menuCourse	ToolStripMenuItem	Text：课程信息管理
AddCourseInfoToolStripMenuItem	ToolStripMenuItem	Text：添加课程信息
MaintainCourseInfoToolStripMenuItem	ToolStripMenuItem	Text：维护课程信息
menuScore	ToolStripMenuItem	Text：成绩信息管理
MaintainScoreInfoToolMenuItem	ToolStripMenuItem	Text：维护成绩信息

2. 功能实现代码

```
using System;
using System.Windows.Forms;
namespace StudentCourseWindowsFormsApp
```

```csharp
    {
        public partial class ManagerMainForm : Form
        {
            public ManagerMainForm()
            {
                InitializeComponent();
            }
            private void ManageSystemInfoToolStripMenuItem_Click(object sender, EventArgs e)
            {
                ManageSystemInfo form = new ManageSystemInfo();
                form.MdiParent = this;
                form.Show();
            }
            private void AddStudentinfoToolStripMenuItem_Click(object sender, EventArgs e)
            {
                AddStudentinfo form = new AddStudentinfo();
                form.MdiParent = this;
                form.Show();
            }
            private void MaintainStudentinfoToolStripMenuItem_Click(object sender, EventArgs e)
            {
                MaintainStudentinfo form = new MaintainStudentinfo();
                form.MdiParent = this;
                form.Show();
            }
            private void AddCourseInfoToolStripMenuItem_Click(object sender, EventArgs e)
            {
                AddCourseInfo form = new AddCourseInfo();
                form.MdiParent = this;
                form.Show();
            }
            private void MaintainCourseInfoToolStripMenuItem_Click(object sender, EventArgs e)
            {
                MaintainCourseInfo form = new MaintainCourseInfo();
                form.MdiParent = this;
                form.Show();
            }
            private void MaintainScoreInfoToolStripMenuItem_Click(object sender, EventArgs e)
            {
```

```
                MaintainScoreInfo form = new MaintainScoreInfo();
                form.MdiParent = this;
                form.Show();
            }
        }
    }
```

9.7.7 系统信息管理设计

1. 窗体设计

"系统信息管理"窗体如图 9-10 所示。

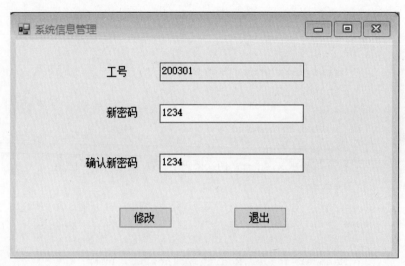

图 9-10 "系统信息管理"窗体

"系统信息管理"窗体和主要控件的属性设置如表 9-12 所示。

表 9-12 "系统信息管理"窗体和主要控件的属性设置

Name	类 型	Text 属性	其 他 属 性
ManageSystemInfo	Form	系统信息管理	
numTxt	TextBox		ReadOnly：True
pwdTxt	TextBox		
pwd2Txt	TextBox		
modifyBtn	Button	修改	
exitBtn	Button	退出	

2. 功能实现代码

```
using System;
using System.Windows.Forms;
using BLL;
namespace StudentCourseWindowsFormsApp
```

```csharp
{
    public partial class ManageSystemInfo : Form
    {
        public ManageSystemInfo()
        {
            InitializeComponent();
        }
        private void modifyBtn_Click(object sender, EventArgs e)
        {
            string num = numTxt.Text;
            string newPwd = pwdTxt.Text;
            string newPwd2 = pwd2Txt.Text;
            if (newPwd == "")
            {
                MessageBox.Show("密码不能为空!");
                return;
            }
            if (newPwd != newPwd2)
            {
                MessageBox.Show("两次密码不一致,请重新输入!");
                return;
            }
            else
            {
                ManagerManager.ModifyPwd(newPwd, num);
                MessageBox.Show("密码修改成功!");
            }
        }
        private void exitBtn_Click(object sender, EventArgs e)
        {
            this.Close();
        }
        private void ManageSystemInfo_Load(object sender, EventArgs e)
        {
            numTxt.Text = RoleManager.curMgr.MgrNum;
        }
    }
}
```

9.7.8 添加学生信息设计

1. 窗体设计

"添加学生信息"窗体如图 9-11 所示。

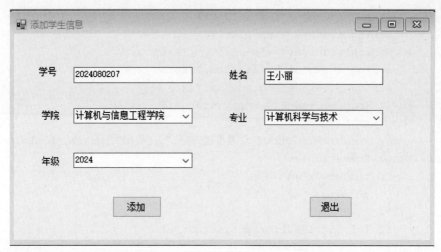

图 9-11 "添加学生信息"窗体

"添加学生信息"窗体和主要控件的属性设置如表 9-13 所示。

表 9-13 "添加学生信息"窗体和主要控件的属性设置

Name	类 型	Text 属性	其 他 属 性
AddStudentinfo	Form	添加学生信息	
numTxt	TextBox		
nameTxt	TextBox		
departCom	ComboBox		
majorCom	ComboBox		
gradeCom	ComboBox		
addBtn	Button	添加	
exitBtn	Button	退出	

2. 功能实现代码

```
using System;
using System.Data;
using System.Windows.Forms;
using Model;
using BLL;
namespace StudentCourseWindowsFormsApp
{
    public partial class AddStudentinfo : Form
    {
        public AddStudentinfo()
        {
            InitializeComponent();
        }
        private void addBtn_Click(object sender, EventArgs e)
        {
```

```csharp
            string stuNum = numTxt.Text;
            string stuName = nameTxt.Text;
            string stuDepart = departCom.Text;
            string stuMajor = majorCom.Text;
            string stuGrade = gradeCom.Text;
            if (string.IsNullOrEmpty(stuNum))
            {
                MessageBox.Show("学号不能为空!", "提示", MessageBoxButtons.OK, MessageBoxIcon.Information);
                numTxt.Focus();
                return;
            }
            if (string.IsNullOrEmpty(stuName))
            {
                MessageBox.Show("姓名不能为空!", "提示", MessageBoxButtons.OK, MessageBoxIcon.Information);
                nameTxt.Focus();
                return;
            }
            if (string.IsNullOrEmpty(stuDepart))
            {
                MessageBox.Show("学院不能为空!", "提示", MessageBoxButtons.OK, MessageBoxIcon.Information);
                return;
            }
            if (string.IsNullOrEmpty(stuMajor))
            {
                MessageBox.Show("专业不能为空!", "提示", MessageBoxButtons.OK, MessageBoxIcon.Information);
                return;
            }
            if (string.IsNullOrEmpty(stuGrade))
            {
                MessageBox.Show("年级不能为空!", "提示", MessageBoxButtons.OK, MessageBoxIcon.Information);
                return;
            }
            Student student = new Student(stuNum, stuName, "123456", stuDepart, stuMajor, stuGrade);
            DataTable dt = StudentManager.GetStudent(stuNum);
            if (dt.Rows.Count != 0)
            {
                MessageBox.Show("学号已存在!");
                return;
            }
            if (StudentManager.AddStudent(student))
```

```csharp
            {
                MessageBox.Show("添加成功!");
            }
            else
            {
                MessageBox.Show("添加失败!");
            }
        }
        private void exitBtn_Click(object sender, EventArgs e)
        {
            this.Close();
        }
        private void departCom_SelectedIndexChanged(object sender, EventArgs e)
        {
            if (departCom.Text == "计算机与信息工程学院")
            {
                majorCom.Text = "";
                majorCom.Items.Clear();
                majorCom.Items.Add("数据科学与大数据技术");
                majorCom.Items.Add("软件工程");
                majorCom.Items.Add("物联网工程");
                majorCom.Items.Add("计算机科学与技术");
            }
            else if (departCom.Text == "矿业工程学院")
            {
                majorCom.Text = "";
                majorCom.Items.Clear();
                majorCom.Items.Add("采矿工程");
                majorCom.Items.Add("矿物加工工程");
                majorCom.Items.Add("资源勘查工程");
                majorCom.Items.Add("测绘工程");
                majorCom.Items.Add("人文地理与城乡规划");
            }
            else if (departCom.Text == "电气与控制工程学院")
            {
                majorCom.Text = "";
                majorCom.Items.Clear();
                majorCom.Items.Add("电气工程及其自动化");
                majorCom.Items.Add("自动化");
                majorCom.Items.Add("测控技术与仪器");
                majorCom.Items.Add("电气工程与智能控制");
            }
        }
    }
}
```

9.7.9 维护学生信息设计

1. 窗体设计

"维护学生信息"窗体如图 9-12 所示。

图 9-12 "维护学生信息"窗体

"维护学生信息"窗体和主要控件的属性设置如表 9-14 所示。

表 9-14 "维护学生信息"窗体和主要控件的属性设置

Name	类 型	Text 属性	其 他 属 性
MaintainStudentinfo	Form	维护学生信息	
groupBox1	GroupBox	查询学生	
groupBox1	GroupBox	详细信息	
searchNumTxt	TextBox		
searchBtn	Button	查询	
clearBtn	Button	清除	
numTxt	TextBox		ReadOnly：True
nameTxt	TextBox		
departCom	ComboBox		
majorCom	ComboBox		
gradeCom	ComboBox		
modifyBtn	Button	修改	
deleteBtn	Button	删除	
exitBtn	Button	退出	

2. 功能实现代码

```csharp
using System;
using System.Data;
using System.Windows.Forms;
using BLL;
using Model;
namespace StudentCourseWindowsFormsApp
{
    public partial class MaintainStudentinfo : Form
    {
        public MaintainStudentinfo()
        {
            InitializeComponent();
        }
        public bool CheckData()
        {
            if (nameTxt.Text == "")
            {
                MessageBox.Show("姓名不能为空!");
                nameTxt.Focus();
                return false;
            }
            if (departCom.Text == "")
            {
                MessageBox.Show("学院不能为空!");
                return false;
            }
            if (majorCom.Text == "")
            {
                MessageBox.Show("专业不能为空!");
                return false;
            }
            if (gradeCom.Text == "")
            {
                MessageBox.Show("年级不能为空!");
                return false;
            }
            return true;
        }
        private void modifyBtn_Click(object sender, EventArgs e)
        {
            Student student = new Student();
            student.StuName = nameTxt.Text;
            student.StuDepart = departCom.Text;
            student.StuMajor = majorCom.Text;
            student.StuGrade = gradeCom.Text;
            student.StuNum = numTxt.Text;
```

```csharp
            if (CheckData())
            {
                if (StudentManager.UpdateStudent(student))
                {
                    MessageBox.Show("修改成功!");
                }
                else
                {
                    MessageBox.Show("修改失败!");
                }
            }
        }
        private void deleteBtn_Click(object sender, EventArgs e)
        {
            if (MessageBox.Show("是否确认删除该学生信息?", "提示", MessageBoxButtons.OKCancel, MessageBoxIcon.Information) == DialogResult.OK)
            {
                Student student = new Student();
                student.StuNum = numTxt.Text;
                if (StudentManager.DeleteStudent(student))
                {
                    MessageBox.Show("成功删除该学生信息!");
                    numTxt.Text = "";
                    nameTxt.Text = "";
                    departCom.Text = "";
                    majorCom.Text = "";
                    gradeCom.Text = "";
                }
                else
                {
                    MessageBox.Show("删除失败!");
                }
            }
        }
        private void exitBtn_Click(object sender, EventArgs e)
        {
            this.Close();
        }
        private void serchBtn_Click(object sender, EventArgs e)
        {
            string stuNum = searchNumTxt.Text;
            if (string.IsNullOrEmpty(stuNum))
            {
                MessageBox.Show("学号不能为空!");
                searchNumTxt.Focus();
                return;
            }
```

```csharp
    DataTable dt = StudentManager.GetStudent(stuNum);
    if (dt.Rows.Count != 0)
    {
        numTxt.Text = dt.Rows[0][0].ToString();
        nameTxt.Text = dt.Rows[0][1].ToString();
        departCom.Text = (dt.Rows[0][3].ToString());
        majorCom.Text = (dt.Rows[0][4]).ToString();
        gradeCom.Text = (dt.Rows[0][5].ToString());
    }
    else
    {
        MessageBox.Show("未查询到该学生,请确认是否输入正确学号!");
        return;
    }
}
private void clearBtn_Click(object sender, EventArgs e)
{
    searchNumTxt.Text = "";
}
private void departCom_SelectedIndexChanged(object sender, EventArgs e)
{
    if (departCom.Text == "计算机与信息工程学院")
    {
        majorCom.Text = "";
        majorCom.Items.Clear();
        majorCom.Items.Add("数据科学与大数据技术");
        majorCom.Items.Add("软件工程");
        majorCom.Items.Add("物联网工程");
        majorCom.Items.Add("计算机科学与技术");
    }
    else if (departCom.Text == "矿业工程学院")
    {
        majorCom.Text = "";
        majorCom.Items.Clear();
        majorCom.Items.Add("采矿工程");
        majorCom.Items.Add("矿物加工工程");
        majorCom.Items.Add("资源勘查工程");
        majorCom.Items.Add("测绘工程");
        majorCom.Items.Add("人文地理与城乡规划");
    }
    else if (departCom.Text == "电气与控制工程学院")
    {
        majorCom.Text = "";
        majorCom.Items.Clear();
        majorCom.Items.Add("电气工程及其自动化");
        majorCom.Items.Add("自动化");
        majorCom.Items.Add("测控技术与仪器");
```

```
                majorCom.Items.Add("电气工程与智能控制");
            }
        }
    }
}
```

9.7.10 添加课程信息设计

1. 窗体设计

"添加课程信息"窗体如图 9-13 所示。

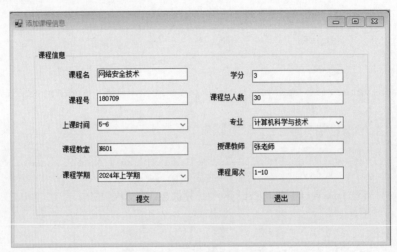

图 9-13 "添加课程信息"窗体

"添加课程信息"窗体和主要控件的属性设置如表 9-15 所示。

表 9-15 "添加课程信息"窗体和主要控件的属性设置

Name	类型	Text 属性	其他属性
AddCourseInfo	Form	添加课程信息	
groupBox1	GroupBox	课程信息	
courseNameTxt	TextBox		
courseCreditTxt	TextBox		
courseNumTxt	TextBox		
coursePersonTxt	TextBox		
courseTimeCom	ComboBox		
courseMajorCom	ComboBox		
courseRoomTxt	TextBox		
courseTeacherTxt	TextBox		
courseTermCom	ComboBox		
courseWeeksTxt	TextBox		
submitBtn	Button	提交	
exitBtn	Button	退出	

2. 功能实现代码

```csharp
using System;
using System.Windows.Forms;
using BLL;
using Model;
namespace StudentCourseWindowsFormsApp
{
    public partial class AddCourseInfo : Form
    {
        public AddCourseInfo()
        {
            InitializeComponent();
        }
        private bool CheckData()
        {
            if (courseNameTxt.Text == "")
            {
                MessageBox.Show("课程名不能为空!");
                courseNameTxt.Focus();
                return false;
            }
            if (courseNumTxt.Text == "")
            {
                MessageBox.Show("课程号不能为空!");
                courseNumTxt.Focus();
                return false;
            }
            else if (!(int.TryParse(courseNumTxt.Text, out int result)))
            {
                MessageBox.Show("课程号需要为整数!");
                courseNumTxt.Focus();
                return false;
            }
            if (courseTermCom.Text == "")
            {
                MessageBox.Show("课程学期不能为空!");
                courseTermCom.Focus();
                return false;
            }
            if (courseTeacherTxt.Text == "")
            {
                MessageBox.Show("授课教师不能为空!");
                courseTeacherTxt.Focus();
                return false;
            }
            if (courseRoomTxt.Text == "")
```

```csharp
        {
            MessageBox.Show("课程教室不能为空!");
            courseRoomTxt.Focus();
            return false;
        }
        if (coursePersonTxt.Text == "")
        {
            MessageBox.Show("课程总人数不能为空!");
            coursePersonTxt.Focus();
            return false;
        }
        else if (!(Convert.ToInt32(coursePersonTxt.Text) <= 30))
        {
            MessageBox.Show("课程总人数要在30人以内!");
            coursePersonTxt.Focus();
            return false;
        }
        if (courseCreditTxt.Text == "")
        {
            MessageBox.Show("课程学分不能为空!");
            courseCreditTxt.Focus();
            return false;
        }
        if (courseWeeksTxt.Text == "")
        {
            MessageBox.Show("课程周次不能为空!");
            courseWeeksTxt.Focus();
            return false;
        }
        if (courseTimeCom.Text == "")
        {
            MessageBox.Show("上课时间不能为空!");
            courseTimeCom.Focus();
            return false;
        }
        if (courseMajorCom.Text == "")
        {
            MessageBox.Show("专业不能为空!");
            courseMajorCom.Focus();
            return false;
        }
        return true;
    }
    private void submitBtn_Click(object sender, EventArgs e)
    {
        if (CheckData() != true)
```

```csharp
        {
            return;
        }
        string courseNum = courseNumTxt.Text;                    //课程号
        string courseName = courseNameTxt.Text;                  //课程名
        string courseTerm = courseTermCom.SelectedItem.ToString();
                                                                 //课程学期
        string courseTeacher = courseTeacherTxt.Text;            //授课老师
        string courseRoom = courseRoomTxt.Text;                  //课程教室
        string coursePerson = coursePersonTxt.Text;              //课程总人数
        string courseCredit = courseCreditTxt.Text;              //课程学分
        string courseWeeks = courseWeeksTxt.Text;                //课程周次
        string courseTime = courseTimeCom.SelectedItem.ToString();
                                                                 //课程时间
        string courseMajor = courseMajorCom.Text;                //专业
        if (CourseManager.SearchCourse(courseNum).Rows.Count != 0)
        {
            MessageBox.Show("课程号已存在!");
            return;
        }
        Course course = new Course(courseNum, courseName, courseTerm,
courseTeacher, courseRoom, coursePerson, courseCredit, courseWeeks, courseTime,
courseMajor);
        if (CourseManager.AddCourse(course))
        {
            MessageBox.Show("添加课程成功!");
        }
        else
        {
            MessageBox.Show("添加课程失败!");
        }
    }
    private void exitBtn_Click(object sender, EventArgs e)
    {
        this.Close();
    }
    private void AddCourseInfo_Load(object sender, EventArgs e)
    {
        courseMajorCom.Items.Add("数据科学与大数据技术");
        courseMajorCom.Items.Add("软件工程");
        courseMajorCom.Items.Add("物联网工程");
        courseMajorCom.Items.Add("计算机科学与技术");
        courseMajorCom.Items.Add("采矿工程");
        courseMajorCom.Items.Add("矿物加工工程");
        courseMajorCom.Items.Add("资源勘查工程");
        courseMajorCom.Items.Add("测绘工程");
```

```
            courseMajorCom.Items.Add("人文地理与城乡规划");
            courseMajorCom.Items.Add("电气工程及其自动化");
            courseMajorCom.Items.Add("自动化");
            courseMajorCom.Items.Add("测控技术与仪器");
            courseMajorCom.Items.Add("电气工程与智能控制");
            courseTimeCom.Items.Add("1-2");
            courseTimeCom.Items.Add("3-4");
            courseTimeCom.Items.Add("5-6");
            courseTimeCom.Items.Add("7-8");
            courseTermCom.Items.Add("2020年上学期");
            courseTermCom.Items.Add("2020年下学期");
            courseTermCom.Items.Add("2021年上学期");
            courseTermCom.Items.Add("2021年下学期");
            courseTermCom.Items.Add("2022年上学期");
            courseTermCom.Items.Add("2022年下学期");
            courseTermCom.Items.Add("2023年上学期");
            courseTermCom.Items.Add("2023年下学期");
            courseTermCom.Items.Add("2024年上学期");
            courseTermCom.Items.Add("2024年下学期");
            courseTermCom.Items.Add("2025年上学期");
            courseTermCom.Items.Add("2026年下学期");
        }
    }
}
```

9.7.11 维护课程信息设计

1. 窗体设计

"维护课程信息"窗体如图 9-14 所示。

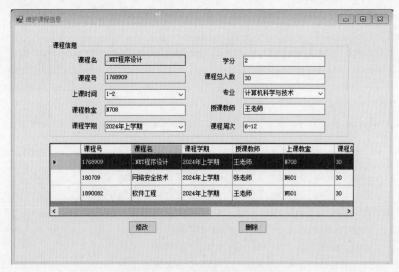

图 9-14 "维护课程信息"窗体

"维护课程信息"窗体和主要控件的属性设置如表 9-16 所示。

表 9-16 "维护课程信息"窗体和主要控件的属性设置

Name	类　型	Text 属性	其 他 属 性
MaintainCourseInfo	Form	维护课程信息	
groupBox1	GroupBox	课程信息	
courseNameTxt	TextBox		ReadOnly：True
courseCreditTxt	TextBox		
courseNumTxt	TextBox		ReadOnly：True
coursePersonTxt	TextBox		
courseTimeCom	ComboBox		
courseMajorCom	ComboBox		
courseRoomTxt	TextBox		
courseTeacherTxt	TextBox		
courseTermCom	ComboBox		
courseWeeksTxt	TextBox		
courseDgv	DataGridView		
modifyBtn	Button	修改	
deleteBtn	Button	删除	

2. 功能实现代码

```
using System;
using System.Data;
using System.Windows.Forms;
using BLL;
using Model;
namespace StudentCourseWindowsFormsApp
{
    public partial class MaintainCourseInfo : Form
    {
        public MaintainCourseInfo()
        {
            InitializeComponent();
        }
        private void MaintainCourseInfo_Load(object sender, EventArgs e)
        {
            courseMajorCom.Items.Add("数据科学与大数据技术");
            courseMajorCom.Items.Add("软件工程");
            courseMajorCom.Items.Add("物联网工程");
            courseMajorCom.Items.Add("计算机科学与技术");
            courseMajorCom.Items.Add("采矿工程");
            courseMajorCom.Items.Add("矿物加工工程");
            courseMajorCom.Items.Add("资源勘查工程");
            courseMajorCom.Items.Add("测绘工程");
```

```csharp
            courseMajorCom.Items.Add("人文地理与城乡规划");
            courseMajorCom.Items.Add("电气工程及其自动化");
            courseMajorCom.Items.Add("自动化");
            courseMajorCom.Items.Add("测控技术与仪器");
            courseMajorCom.Items.Add("电气工程与智能控制");
            courseTimeCom.Items.Add("1-2");
            courseTimeCom.Items.Add("3-4");
            courseTimeCom.Items.Add("5-6");
            courseTimeCom.Items.Add("7-8");
            courseTermCom.Items.Add("2020年上学期");
            courseTermCom.Items.Add("2020年下学期");
            courseTermCom.Items.Add("2021年上学期");
            courseTermCom.Items.Add("2021年下学期");
            courseTermCom.Items.Add("2022年上学期");
            courseTermCom.Items.Add("2022年下学期");
            courseTermCom.Items.Add("2023年上学期");
            courseTermCom.Items.Add("2023年下学期");
            courseTermCom.Items.Add("2024年上学期");
            courseTermCom.Items.Add("2024年下学期");
            courseTermCom.Items.Add("2025年上学期");
            courseTermCom.Items.Add("2026年下学期");
            DataTable dt = CourseManager.GetCourseInfo();
            if (dt.Rows.Count != 0)
            {
                courseDgv.DataSource = dt;
            }
            else
            {
                MessageBox.Show("没有课程信息!");
                return;
            }
        }
        private bool CheckData()
        {
            if (courseNameTxt.Text == "")
            {
                MessageBox.Show("课程名不能为空!");
                courseNameTxt.Focus();
                return false;
            }
            if (courseNumTxt.Text == "")
            {
                MessageBox.Show("课程号不能为空!");
                courseNumTxt.Focus();
                return false;
            }
```

```csharp
else if (!(int.TryParse(courseNumTxt.Text, out int result)))
{
    MessageBox.Show("课程号需要为整数!");
    courseNumTxt.Focus();
    return false;
}
if (courseTermCom.Text == "")
{
    MessageBox.Show("课程学期不能为空!");
    courseTermCom.Focus();
    return false;
}
if (courseTeacherTxt.Text == "")
{
    MessageBox.Show("授课教师不能为空!");
    courseTeacherTxt.Focus();
    return false;
}
if (courseRoomTxt.Text == "")
{
    MessageBox.Show("课程教室不能为空!");
    courseRoomTxt.Focus();
    return false;
}
if (coursePersonTxt.Text == "")
{
    MessageBox.Show("课程总人数不能为空!");
    coursePersonTxt.Focus();
    return false;
}
else if (!(Convert.ToInt32(coursePersonTxt.Text) <= 30))
{
    MessageBox.Show("课程总人数要在30人以内!");
    coursePersonTxt.Focus();
    return false;
}
if (courseCreditTxt.Text == "")
{
    MessageBox.Show("课程学分不能为空!");
    courseCreditTxt.Focus();
    return false;
}
if (courseWeeksTxt.Text == "")
{
    MessageBox.Show("课程周次不能为空!");
```

```csharp
            courseWeeksTxt.Focus();
            return false;
        }
        if (courseTimeCom.Text == "")
        {
            MessageBox.Show("上课时间不能为空!");
            courseTimeCom.Focus();
            return false;
        }
        if (courseMajorCom.Text == "")
        {
            MessageBox.Show("专业不能为空!");
            courseMajorCom.Focus();
            return false;
        }
        return true;
    }
    private void modifyBtn_Click(object sender, EventArgs e)
    {
        Course course = new Course();
        course.CourseMajor = courseMajorCom.Text;
        course.CourseWeeks = courseWeeksTxt.Text;
        course.CoursePerson = coursePersonTxt.Text;
        course.CourseNum = courseNumTxt.Text;
        course.CourseCredit = courseCreditTxt.Text;
        course.CourseName = courseNameTxt.Text;
        course.CourseRoom = courseRoomTxt.Text;
        course.CourseTeacher = courseTeacherTxt.Text;
        course.CourseTerm = courseTermCom.Text;
        course.CourseTime = courseTimeCom.Text;
        if (CheckData())
        {
            if (CourseManager.UpdateCourse(course))
            {
                MessageBox.Show("课程信息修改成功!");
                courseDgv.DataSource = CourseManager.GetCourseInfo();
            }
            else
            {
                MessageBox.Show("课程信息修改失败!");
            }
        }
    }
    private void deleteBtn_Click(object sender, EventArgs e)
    {
```

```csharp
            if (MessageBox.Show("是否确认删除该课程信息？", "提示",
MessageBoxButtons.OKCancel, MessageBoxIcon.Information) == DialogResult.OK)
            {
                Course course = new Course();
                if (courseNumTxt.Text != "" || courseNameTxt.Text != "")
                {
                    course.CourseNum = courseNumTxt.Text;
                    course.CourseName = courseNameTxt.Text;
                }
                else
                {
                    MessageBox.Show("请输入需要删除的课程信息!");
                    return;
                }
                if (CourseManager.DeleteCourse(course))
                {
                    MessageBox.Show("成功删除该课程信息!");
                    DataTable dt = CourseManager.GetCourseInfo();
                    courseDgv.DataSource = dt;
                    //清空信息
                    courseMajorCom.Text = "";
                    courseWeeksTxt.Text = "";
                    coursePersonTxt.Text = "";
                    courseNumTxt.Text = "";
                    courseCreditTxt.Text = "";
                    courseNameTxt.Text = "";
                    courseRoomTxt.Text = "";
                    courseTeacherTxt.Text = "";
                    courseTermCom.Text = "";
                    courseTimeCom.Text = "";
                }
                else
                {
                    MessageBox.Show("删除失败!");
                }
            }
        }
        private void courseDgv_CellContentClick(object sender,
DataGridViewCellEventArgs e)
        {
            if (courseDgv.SelectedRows.Count == 1)
            {
                courseNameTxt.Text = Convert.ToString
(courseDgv.SelectedRows[0].Cells["课程名"].Value.ToString());
                courseNumTxt.Text = Convert.ToString
(courseDgv.SelectedRows[0].Cells["课程号"].Value.ToString());
```

```
                    courseTermCom.Text = Convert.ToString
(courseDgv.SelectedRows[0].Cells["课程学期"].Value.ToString());
                    courseMajorCom.Text = Convert.ToString
(courseDgv.SelectedRows[0].Cells["专业"].Value.ToString());
                    courseTeacherTxt.Text = Convert.ToString
(courseDgv.SelectedRows[0].Cells["授课教师"].Value.ToString());
                    courseRoomTxt.Text = Convert.ToString
(courseDgv.SelectedRows[0].Cells["上课教室"].Value.ToString());
                    coursePersonTxt.Text = Convert.ToString
(courseDgv.SelectedRows[0].Cells["课程总人数"].Value.ToString());
                    courseCreditTxt.Text = Convert.ToString
(courseDgv.SelectedRows[0].Cells["学分"].Value.ToString());
                    courseWeeksTxt.Text = Convert.ToString
(courseDgv.SelectedRows[0].Cells["课程周次"].Value.ToString());
                    courseTimeCom.Text = Convert.ToString
(courseDgv.SelectedRows[0].Cells["上课时间"].Value.ToString());
                }
            }
        }
    }
```

9.7.12 维护成绩信息设计

1. 窗体设计

"维护成绩信息"窗体如图 9-15 所示。

图 9-15 "维护成绩信息"窗体

"维护成绩信息"窗体和主要控件的属性设置如表 9-17 所示。

表 9-17 "维护成绩信息"窗体和主要控件的属性设置

Name	类 型	Text 属性	其 他 属 性
MaintainScoreInfo	Form	维护成绩信息	
courseTermCOm	ComboBox		
stuNumTxt	TextBox		
courseNameCom	ComboBox		
searchBtn	Button	查询	
scoreDgv	DataGridView		
scoreTxt	TextBox		
submitBtn	Button	提交	

2. 功能实现代码

```
using System;
using System.Collections.Generic;
using System.Data;
using System.Windows.Forms;
using BLL;
using MySql.Data.MySqlClient;
namespace StudentCourseWindowsFormsApp
{
    public partial class MaintainScoreInfo : Form
    {
        public MaintainScoreInfo()
        {
            InitializeComponent();
        }
        private void searchBtn_Click(object sender, EventArgs e)
        {
            //获取用户输入
            string courseTerm = courseTermCom.Text;
            string courseName = courseNameCom.Text;
            string stuNum = stuNumTxt.Text;
            //查询学生成绩表
            DataTable dt = ScoreManager.GetScoreInfo(stuNum, courseName, courseTerm);
            scoreDgv.DataSource = dt;
        }
        //验证提交分数
        public bool CheckScore()
        {
            bool flag = true;
            if (scoreTxt.Text == "")
            {
                MessageBox.Show("分数不能为空!");
```

```csharp
            flag = false;
        }
        else if (Convert.ToInt32(scoreTxt.Text) < 0 || Convert
.ToInt32(scoreTxt.Text) > 100)
        {
            MessageBox.Show("分数需要在0到100之间，请重新输入！");
            flag = false;
        }
        return flag;
    }
    private void submitBtn_Click(object sender, EventArgs e)
    {
        //验证用户输入分数
        if (CheckScore() == false)
        {
            return;
        }
        //获取用户输入
        int courseScore = Convert.ToInt32(scoreTxt.Text);
        string courseName = courseNameCom.Text;
        string stuNum = stuNumTxt.Text;
        string courseTerm = courseTermCom.Text;
        //提交或修改成绩
        if (ScoreManager.UpdateScore(courseName, stuNum, courseScore))
        {
            MessageBox.Show("提交成功！");
        }
        else
        {
            MessageBox.Show("提交失败,请检查信息是否正确！");
        }
        scoreDgv.DataSource = ScoreManager.GetScoreInfo(stuNum,
courseName, courseTerm);
    }
    //填充课程名与课程学期的下拉列表框
    private void MaintainScoreInfo_Load(object sender, EventArgs e)
    {
        List<string> name = new List<string>();
        List<string> term = new List<string>();
        MySqlDataReader reader = ScoreManager.GetTermAndName();
        while (reader.Read())
        {
            name.Add(reader.GetString(0));
            term.Add(reader.GetString(1));
        }
```

```csharp
            courseNameCom.DataSource = name;
            courseTermCom.DataSource = term;
            reader.Close();
            courseNameCom.SelectedIndex = -1;
            courseTermCom.SelectedIndex = -1;
        }
        private void scoreDgv_CellContentClick(object sender, DataGridViewCellEventArgs e)
        {
            if (scoreDgv.SelectedRows.Count == 1)
            {
                scoreTxt.Text = Convert.ToString(scoreDgv.SelectedRows[0].Cells["成绩"].Value);
                stuNumTxt.Text = Convert.ToString(scoreDgv.SelectedRows[0].Cells["学号"].Value);
                courseNameCom.Text = Convert.ToString(scoreDgv.SelectedRows[0].Cells["课程名"].Value);
                courseTermCom.Text = Convert.ToString(scoreDgv.SelectedRows[0].Cells["学期"].Value);
            }
        }
    }
}
```

参 考 文 献

[1] 江红，余青松. C#程序设计教程[M]. 4版. 北京：清华大学出版社，2023.
[2] 郭基凤，高亮，韩玉民，等. 基于C#的管理信息系统开发[M]. 3版. 北京：清华大学出版社，2023.
[3] 江红，余青松. C#程序设计实验指导与习题测试[M]. 4版. 北京：清华大学出版社，2023.
[4] 草宇，许高峰，王佳丽. C#程序设计与编程案例（微课视频版）[M]. 北京：清华大学出版社，2022.
[5] 黄兴荣，李昌领，张廷秀，等. C#边做边学[M]. 北京：清华大学出版社，2021.